WHEN DO FISH SLEEP?

ALSO BY DAVID FELDMAN

Do Elephants Jump?

How Do Astronauts Scratch an Itch?

What Are Hyenas Laughing at, Anyway?

How to Win at Just About Anything

How Does Aspirin Find a Headache?

Are Lobsters Ambidextrous?

  (Originally published as *When Did Wild Poodles Roam the Earth?*)

Do Penguins Have Knees?

Why Do Dogs Have Wet Noses?

Who Put the Butter in Butterfly?

Why Do Clocks Run Clockwise?

Why Don't Cats Like to Swim?

  (Originally published as *Imponderables*)

WHEN DO FISH SLEEP?
An Imponderables® Book

David Feldman

Illustrated by Kassie Schwan

Collins

An Imprint of HarperCollinsPublishers

A hardcover edition of this book was published in 1989 by Harper & Row, Publishers, Inc.

HarperCollins books may be purchased for educational, business, or sales promotional use. For information please write: Special Markets Department, HarperCollins Publishers Inc., 10 East 53rd Street, New York, NY 10022.

First Perennial Currents edition published 2005.

Library of Congress Cataloging-in-Publication Data
Feldman, David.
    When do fish sleep? : an imponderables book / David Feldman ; illustrated by Kassie Schwan.
    p. cm.
    Includes index.
    ISBN 0-06-074093-0 (pbk.)
    1. Questions and answers. 2. Questions and answers—Humor. I. Title.

AG195.F458 2005
031.02—dc22

                                                                2004059635

        07 08 09 RRD H 10 9

For Phil and Gilda Feldman

Preface

Imponderables are mysteries that can't be answered by numbers, measurements, or a trip to the reference section of your library. If you worry about why the carbons on airplane tickets are red, or why tennis balls are fuzzy, or why yawning is contagious, you have been struck by the dread malady of Imponderability.

When we wrote the first volume of *Imponderables*, we weren't sure that there were others like us, who were committed to cogitating about the everyday mysteries of life. We needn't have worried. Most of the Imponderables in this book were submitted by readers of the first two volumes of Imponderables.

In *Why Do Clocks Run Clockwise?*, we introduced a new section, Frustables (short for "frustrating Imponderables") and asked for your help in solving them. Your response was terrific, but we don't want you to get complacent. We've got ten new Imponderables that we haven't been able to solve.

And because so many readers offered corrections and caustically constructive comments, we've added a letter section—we couldn't shut you up anymore even if we wanted to.

Would you like to win a free copy of the next volume of *Imponderables?* If you are the first to submit an Imponderable that we use in the next book, you will not only have the relief of finally having the answer to your mystery, but also a free, autographed copy of the book (along with, of course, an acknowledgment).

The last page of the book tells you how to get in touch with us. But for now, sit back and enjoy. You are about to enter the wonderful world of Imponderability.

Imponderables

Why Do Roosters Crow in the Morning?

Because there are humans around to be awakened, of course. Does anyone really believe that roosters crow when they are by themselves? Nah! Actually, they speak perfectly good English.

Ornithologists don't buy our common-sense answer. They insist that crowing "maps territory" (a euphemism for "Get the hell out of my way and don't mess with my women—this is my coop"). In the spirit of fair play, we'll give the last word to one of those nasty ornithologist types (but don't believe a word she says), Janet Hinshaw, of the Wilson Ornithological Society:

> Most of the crowing takes place in the morning, as does most singing, because that is when the birds are most active, and most of the territorial advertising takes place then. Many of the other vocalizations heard throughout the day are for other types of communication, including flocking calls, which serve to keep members of a flock together and in touch if they are out of sight from one another.

Submitted by Rowena Nocom of North Hollywood, California.

Why Do Many Hotels and Motels Fold Over the Last Piece of Toilet Paper in the Bathroom?

This Imponderable was sent in by reader Jane W. Brown in a letter dated May 12, 1986. Jane was clearly a discerning seer of emergent popular culture trends:

> Staying in less than deluxe lodgings has led me to wonder why, and how, the custom of folding under the two outside corners on a roll of bathroom paper was begun. This operation creates a V on the last exposed edge of the tissue. I first noticed this bizarre sight in a LaQuinta Motor Inn. Then I stayed in some Holiday Inns while on a business trip. There, too, the bathroom paper had been tediously tucked in on the outside edges, the large V standing out, begging for attention. Recently, I upgraded my accommodations and spent several nights in a Marriott and an Intercontinental. Right: the bathroom paper was also arranged in this contorted fashion. Why?

Jane, enterprisingly, included an audiovisual aid along with her letter, as if to prove she wasn't crazy: a specimen of the mysterious V toilet paper. Since Jane wrote her letter, the folded toilet paper trick has run rampant in the lodging industry.

We contacted most of the largest chains of innkeepers in the country and received the same answer from all. Perhaps James P. McCauley, executive director of the International Association of Holiday Inns, stated it best:

> Hotels want to give their guests the confidence that the bathroom has been cleaned since the last guest has used the room. To accomplish this, the maid will fold over the last piece of toilet paper to assure that no one has used the toilet paper since the room was cleaned. It is subtle but effective.

Maybe too subtle for us. Call us sentimental old fools, but we still like the old "Sanitized for Your Protection" strips across the toilet seat.

Submitted by Jane W. Brown of Giddings, Texas.

DAVID FELDMAN

Why Do Gas Gauges in Automobiles Take an Eternity to Go from Registering Full to Half-full, and Then Drop to Empty in the Speed of Light?

On a long trek down our illustrious interstate highway system, we will do anything to alleviate boredom. The roadway equivalent of reading cereal boxes at breakfast is obsessing about odometers and fuel gauges.

Nothing is more dispiriting after a fill-up at the service station than traveling sixty miles and watching the gas gauge stand still. Although part of us longs to believe that our car is registering phenomenal mileage records, the other part of us wants the gauge to move to prove to ourselves that we are actually making decent time and have not, through some kind of *Twilight Zone* alternate reality, actually been riding on a treadmill for the last hour. Our gas gauge becomes the arbiter of our progress. Even when the needle starts to move, and the gauge registers three-quarters' full, we sometimes feel as if we have been traveling for days.

How nice it would be to have a gauge move steadily down toward empty. Just as we are about to give in to despair, though, after the gauge hits half-full, the needle starts darting toward empty as if it had just discovered the principle of gravity. Whereas it seemed that we had to pass time zones before the needle would move to the left at all, suddenly we are afraid that we are going to run out of gas. Where is that next rest station?

There must be a better way. Why don't fuel gauges actually register what proportion of the tank is filled with gasoline? The automakers and gauge manufacturers are well aware that a "half-full" reading on a gas gauge is really closer to "one-third" full, and they have reasons for preserving this inaccuracy.

The gauge relies upon a sensor in the tank to relay the fuel level. The sensor consists of a float and linkage connected to a variable resistor. The resistance value fluctuates as the float moves up and down.

If a gas tank is filled to capacity, *the liquid is filled higher than the float has the physical ability to rise.* When the float is at the top of its stroke, the gauge will always register as full, *even though the tank can hold more gasoline.* The gauge will register full until this "extra" gasoline is consumed and the float starts its descent in the tank. At the other end of the float's stroke, *the gauge will register as empty when the float can no longer move further downward, even though liquid is present below the float.*

We asked Anthony H. Siegel, of Ametek's U.S. Gauge Division, why sensors aren't developed that can measure the actual status of gasoline more accurately. We learned, much as we expected, that more precise measurements easily could be produced, but the automakers are using the current technology *for our own good:*

> Vehicle makers are very concerned that their customers do not run out of fuel before the gauge reads empty. That could lead to stranded, unhappy motorists, so they compensate in the design of the float/gauge system. Their choice of tolerances and calibration procedures guarantees that slight variations during the manufacturing of these components will always produce a combination of parts which falls on the safe side. The gauge is thus designed to read empty when there is still fuel left.

Tens of millions of motorists have suspected there is fuel left even when the gauge says empty, but few have been brave enough to test the hypothesis. Perhaps there are gallons and gallons of fuel left when the gauge registers empty, and this is all a plot by Stuckey's and Howard Johnson's to make us take unnecessary pit stops on interstates.

Submitted by Jack Belck of Lansing, Michigan.

DAVID FELDMAN

How Is the Caloric Value of Food Measured?

Imponderables is on record as doubting the validity of caloric measurements. It defies belief that the caloric value of vegetables such as potato chips and onion rings, full of nutrients, could possibly be higher than greasy tuna fish or eggplant. Still, with an open mind, we sought to track down the answer to this Imponderable.

Calories are measured by an apparatus called a *calorimeter*. The piece of food to be measured is placed inside a chamber, sealed, and then ignited and burned. The energy released from the food heats water surrounding the chamber. By weighing the amount of water heated, noting the increase in the water's temperature and multiplying the two, the energy capacity of the food can be measured. A calorie is nothing more than the measurement of the ability of a particular nutrient to raise the temperature of one gram of water one degree Centigrade. For example, if ten thousand grams of water (the equivalent of ten liters or ten thousand cubic centimeters) surrounding the chamber is 20 degrees Centigrade before combustion and then is measured at

25 degrees after combustion, the difference in temperature (five degrees) is multiplied by the volume of water (ten thousand grams) to arrive at the caloric value (fifty thousand calories of energy).

If fifty thousand calories sounds like too high a number to describe heating ten liters of water five degrees, your instincts are sound. One calorie is too small a unit of measurement to be of practical use, so the popular press uses "calorie" to describe what the scientists call "Calories," really kilocalories, one thousand times as much energy as the lowercase "calorie."

The calorimeter is a crude but reasonable model for how our body stores and burns energy sources. The calorimeter slightly overstates the number of calories our body can use from each foodstuff. In the calorimeter, foods burn completely, with only some ashes (containing minerals) left in the chamber. In our body, small portions of food are indigestible, and are excreted before they break down to provide energy. The rules of thumb are that two percent of fat, five percent of carbohydrates, and eight percent of proteins will not be converted to energy by the body.

Food scientists have long known the caloric count for each food group. One gram of carbohydrates or proteins equals four calories. One gram of fat contains more than twice the number of calories (nine).

Scientists can easily ascertain the proportion of fat to carbohydrates or proteins, so it might seem that calories could be measured simply by weighing the food. When a food consists exclusively of proteins and carbohydrates, for example, one could simply multiply the weight of the food by four to discover the calorie count.

But complications arise. Certain ingredients in natural or processed foods contain no caloric value whatsoever, such as water, fiber, and minerals. Foods that contain a mixture, say, of water (zero calories), fiber (zero calories), proteins (four calories per gram), fats (nine calories per gram), and carbohydrates (four calories per gram), along with some trace minerals (zero calories), are simply harder to calculate with a scale than a calorimeter.

Submitted by Jill Palmer of Leverett, Massachusetts.

DAVID FELDMAN

Who Put E on Top of the Eye Chart? And Why?

Professor Hermann Snellen, a Dutch professor of ophthalmology, put the E on top of the eye chart in 1862. Although his very first chart was headed by an A, Snellen quickly composed another chart with E on top.

Snellen succeeded Dr. Frans Cornelis Donders as the director of the Netherlands Hospital for Eye Patients. Donders was then the world's foremost authority on geometric optics. Snellen was trying to standardize a test to diagnose visual acuity, to measure how small an image an eye can accept while still detecting the detail of that image. Dr. Donders' complicated formulas were based on three parallel lines; of all the letters of the alphabet, the capital E most closely resembled the lines that Dr. Donders had studied so intensively. Because Donders had earlier determined how the eye perceives the E, Snellen based much of his mathematical work on the fifth letter.

The three horizontal limbs of the E are separated by an

equal amount of white space. In Snellen's original chart, there was a one-to-one ratio between the height and width of the letters, and the gaps and bars were all the same length (in some modern eye charts, the middle bar is shorter).

Louanne Gould, of Cambridge Instruments, says that the E, unlike more open letters like L or U, forces the observer to distinguish between white and black, an important consitituent of good vision. Without this ability, E's begin to look like B's, F's, P's or many other letters.

Of course, Snellen couldn't make an eye chart full of only E's, or else all his patients would have 20-10 vision. But Snellen realized that it was important to use the same letters many times on the eye charts, to insure that the failure of an observer to identify a letter was based on a visual problem rather than the relative difficulty of a set of letters. Ian Bailey, professor of optometry and director of the Low Vision Clinic at the University of California at Berkeley, says that it isn't so important whether an eye chart uses the easiest or most difficult letters. Most eye charts incorporate only ten different letters, ones that have the smallest range of difficulty.

Today, many eye charts do not start with an E—and there is no technical reason why they have to—but most still do. Dr. Stephen C. Miller, of the American Optometric Association, suggests that the desire of optical companies to have a standardized approach to the production of eye charts probably accounts for the preponderance of E charts. And we're happy about it. It's a nice feeling to know that even if our vision is failing us miserably, we'll always get the top row right.

Submitted by Merry Phillips of Menlo Park, California.

Do the Police Really Make Chalk Outlines of Murder Victims at the Scene of the Crime? Why Do They Use Chalk?

As soon as law enforcement officials descend upon a murder scene, a police photographer takes pictures of the corpse, making certain that the deceased's position is established by the photographs. The medical examiner usually wants the body as soon as possible after the murder; the sooner an autopsy is conducted, the more valuable the information the police are likely to obtain.

Right before the body is removed, the police do indeed make an outline of the position of the victim. More often than not the body is outlined in chalk, including a notation of whether the body was found in a prone or supine posture.

A police investigation of a murder can take a long time, too long to maintain the murder site as it appeared after the murder. Forensic specialists cannot rely on photographs alone. Often, the exact position of the victim can be of vital importance in an investigation. By making an outline, the police can return to the murder scene and take measurements which might quash or corroborate a new theory on the case. Outline drawings may also be used in the courtroom to explain wound locations, bullet trajectories, and blood trails.

Herbert H. Buzbee, of the International Association of Coroners and Medical Examiners, told *Imponderables* that chalk is not always used to make outlines. Stick-em paper or string are often used on carpets, for example, where chalk might be obscured by the fabric. Carl Harbaugh, of the International Chiefs of Police, says that many departments once experimented with spray paint to make outlines, but found that paint traces were occasionally found on the victim, confusing the forensic analysis.

The ideal outline ingredient would be one that would show up, stay put, and do no permanent damage to any surface. Unfortunately, no such ingredient exists. Chalk gets high marks for leaving no permanent markings, but is not easily visible on many surfaces. Tape and string (which has to be fastened with tape) have a tendency to mysteriously twist out of shape, especially if they get wet.

None of these flaws in the markers would matter if murder victims were considerate enough to die in sites convenient to the police. Harbaugh says that on a street or highway any kind of outline will do. But what good is a chalk outline on a bed covered with linens and blankets?

Submitted by Pat O'Conner of Forest Hills, New York.

What Do Restaurants that Specialize in Potato Skins Do with the Rest of the Potato? What Do Restaurants that Specialize in Frogs' Legs Do with the Rest of the Frog?

In most restaurants, potato skins are a waste product, served as the casing of a baked potato or not at all. So we assumed that restaurants that specialized in potato skins used the rest of the potato to make mashed potatoes, boiled potatoes, or soups.

Our assumption was correct, but our correspondent mentions that potato skins are often served in bars that do not serve potatoes in any other form. Is it cost-effective for these establishments to serve the skins and dump the potato filling?

Most restaurants that serve potato skins buy the skins *only*, usually in frozen form. Linda Smith, of the National Restaurant Association, sent us a list of the biggest suppliers of potato skins. Most of these companies, not at all coincidentally, also supply restaurants with pre-cut cottage fries, hash browns, and O'Brien potatoes, among others. Ore-Ida isn't about to sell the skin and throw away the potato.

Anyone who has ever dissected a frog in biology class does not want to contemplate the idea of chefs picking apart an entire frog to get at its legs. Suffice it to say that restaurants buy only the legs of frogs. What suppliers of frogs' legs do with the rest of the frog is too gruesome for even us to contemplate.

Submitted by Myrna S. Gordon of Scotch Plains, New Jersey.
Thanks also to Sharon Michele Burke of Menlo Park, California.

If Water Is Heavier than Air, Why Do Clouds Stay Up in the Sky?

What makes you think that clouds aren't dropping? They are. Constantly.

Luckily, cloud drops do not fall at the same velocity as a water balloon. In fact, cloud drops are downright sluggards: They drop at a measly 0.3 centimeters per second. And cloud drops are so tiny, about 0.01 centimeters in diameter, that their descent is not even noticeable to the human eye.

Submitted by Ronald C. Semone of Washington, D.C.

Why Are There More Holes in the Mouthpiece of a Telephone than in the Earpiece?

We just checked the telephone closest to us and were shocked. There are thirty-six holes on our mouthpiece, and a measly seven on the earpiece. What gives?

Tucked underneath the mouthpiece is a tiny transmitter that duplicates our voices, and underneath the earpiece is a receiver. Those old enough to remember telephones that constantly howled will appreciate the problems inherent in having a receiver and transmitter close together enough to produce audible transmission without creating feedback.

Before the handset, deskstand telephones were not portable, and the speaker had to talk into a stationary transmitter. Handsets added convenience to the user but potential pitfalls in transmission. While developing the telephone handset, engineers were aware that it was imperative for the lips of a speaker to be as close as possible to the transmitter. If a caller increases

DAVID FELDMAN

the distance between his lips and the transmitter from half an inch to one inch, the output volume will be reduced by three decibels. According to AT&T, in 1919 more than four thousand measurements of head dimensions were made to determine the proper dimensions of the handset. The goal, of course, was to design a headset that would best cup the ear and bring the transmitter close to the lips.

One of the realities that the Bell engineers faced was that there was no way to force customers to talk directly into the mouthpiece. Watch most people talking on the phone and you will see their ears virtually covered by the receiver. But most people do not hold their mouths as close to the transmitter. This is the real reason why there are usually more holes in the mouthpiece than in the earpiece. The more holes there are, the more sensitive to sound the transmitter is, and the more likely that a mumbled aside will be heard three thousand miles away.

Submitted by Tammy Madill of Millington, Tennessee.

How Do Fish Return to a Lake or Pond that Has Dried Up?

Our correspondent, Michael J. Catalana, rightfully wonders how even a small pond replenishes itself with fish after it has totally dried up. Is there a Johnny Fishseed who roams around the world restocking ponds and lakes with fish?

We contacted several experts on fish to solve this mystery, and they wouldn't answer until we cross-examined you a little bit, Michael. "How carefully did you look at that supposedly dried-up pond?" they wanted to know. Many species, such as the appropriately named mudminnows, can survive in mud. R. Bruce Gebhardt, of the North American Native Fishes Association, suggested that perhaps your eyesight was misdirected: "If there are small pools, fish may be able to hide in mud or weeds while you're standing there looking into the pool." When you leave, they re-emerge. Some tropical fish lay eggs that develop while the pond is dry; when rain comes and the pond is refilled with water, the eggs hatch quickly.

For the sake of argument, Michael, we'll assume that you communed with nature, getting down on your hands and knees to squeeze the mud searching for fish or eggs. You found no evidence of marine life. How can fish appear from out of thin air? We return to R. Bruce Gebhardt for the explanation:

> There are ways in which fish can return to a pond after total elimination. The most common is that most ponds or lakes have outlets and inlets; fish just swim back into the formerly hostile area. They are able to traverse and circumvent small rivulets, waterfalls, and pollution sources with surprising efficiency. If they find a pond with no fish in it, they may stay just because there's a lot of food with no competition for it.

Submitted by Michael J. Catalana of Ben Lomond, California.

Why Do We Call Our Numbering System "Arabic" When Arabs Don't Use Arabic Numbers Themselves?

The first numbering system was probably developed by the Egyptians, but ancient Sumeria, Babylonia, and India used numerals in business transactions. All of the earliest number systems used some variation of 1 to denote one, probably because the numeral resembled a single finger. Historians suggest that our Arabic 2 and 3 are corruptions of two and three slash marks written hurriedly.

Most students in Europe, Australia, and the Americas learn to calculate with Arabic numbers, even though *these numerals were never used by Arabs*. Arabic numbers were actually developed in India, long before the invention of the printing press (probably in the tenth century), but were subsequently translated into Arabic. European merchants who brought back trea-

DAVID FELDMAN

tises to their continent mistakenly assumed that Arabs had invented the system, and proceeded to translate the texts from Arabic.

True Arabic numerals look little like ours. From one to ten, this is how they look:

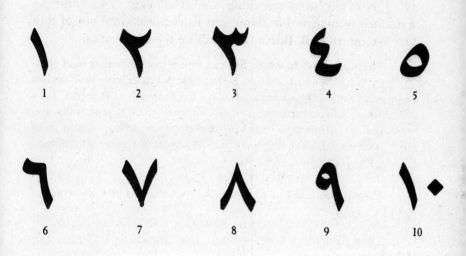

Submitted by Dr. Bruce Carter of Fort Ord, California.

When You Are Driving Your Car at Night and Look Up at the Sky, Why Does It Seem that the Moon Is Following You Around?

If you, like every other literate human being, have read *Why Do Clocks Run Clockwise? and Other Imponderables*, then you know why the moon looks larger on the horizon than up in the sky, even though the moon remains the same size. Clearly, our eyes can play tricks on us.

Without reference points to guide us, the moon doesn't seem to be far away. When you are driving on a highway, the objects closest to your car go whirring by. Barriers dividing the lanes become a blur. You can discern individual houses or trees by the side of the road, but, depending upon your speed, it might be painful to watch them go by. Distant trees and houses move by much more slowly, even though you are driving at the same speed. And distant mountains seem mammoth and motionless.

DAVID FELDMAN

Eventually, as you travel far enough down the highway, you will pass the mountains, and they will appear smaller.

If you think the mountain range off the highway is large or far away, consider the moon, which is 240,000 miles away and bigger than any mountain range (more than 2,100 miles in diameter). We already know that our eyes are playing tricks with our perception of how big and far away the moon is. You would have to be traveling awfully far to make the moon appear to move at all. *Astronomy* editor Jeff Kanipe concludes that without a highway or expanse of landscape to give us reference points "this illusion of nearness coupled with its actual size and distance makes the moon appear to follow us wherever we go."

This phenomenon, much discussed in physics and astronomy textbooks, is called the parallax and is used to determine how the apparent change in the position of an object or heavenly body may be influenced by the changing position of the observer. Astronomers can determine the distance between a body in space and the observer by measuring the magnitude of the parallax effect.

And then again, Elizabeth, maybe the moon really is following you.

Submitted by Elizabeth Bogart of Glenview, Illinois.

When Does a Calf Become a Cow?

The calf's equivalent of a bar mitzvah occurs after it stops nursing, usually at about seven to eight months of age. After they are weaned and/or when they reach twelve months, they are referred to as yearling bulls or yearling calves.

According to Richard L. Spader, executive vice president of the American Angus Association, "calves don't achieve full-fledged bullhood or cowhood until they're in production. We normally refer to a first calf heifer at, say, twenty-four months of age or older, as just that, and after her second calf as a three-year-old, she becomes a cow."

Bulls don't usually reach maturity until they are three. After they wean from their mothers, they are referred to as "yearling bulls," or "two-year-old bulls." Are we now all totally confused?

Submitted by Herbert Kraut of Forest Hills, New York.

When One Has a Cold, Why Does Only One Nostril at a Time Tend to Get Clogged (Even Though *Which* Nostril Gets Clogged Can Change at Any Time)? Come to Think of It, Why Do We Need Two Nostrils in the First Place?

The shifting of clogged nostrils is a protective effort of your nasal reflex system. Although the nose was probably most important to prehistoric man as a smelling organ, modern humans' sense of smell has steadily decreased over time. The nose is now much more important in respiration, breathing in O_2 to the nose, trachea, bronchi, lungs, heart, and blood, and ultimately the exchange of oxygen and carbon dioxide. As rhinologist Dr. Pat Barelli explains:

> A fantastic system of reflexes which originate in the inner nose sends impulses to the heart and indirectly to every cell in the body. These reflexes, coupled with the resistance of the nose, increase the efficiency of the lungs and improve the effectiveness of the heart action.

The most common reason for congested nostril switching is the sleep process. When we sleep, our body functions at a greatly reduced rate. The heart beats slower and the lungs require less air. Rhinologist Dr. Zanzibar notes that patients

> commonly complain that at night when they lie on one side, the dependent side of the nose becomes obstructed and they find it necessary to roll over in bed to make that side open. Then the other side becomes obstructed, and they roll over again.

When the head is turned to one side during sleep, the "upper nose" has the entire load of breathing and can become fatigued. According to Dr. Pat Barelli,

> one nostril doing solo duty can fatigue in as little as one to three hours, and internal pressures cause the sleeper to change his head position to the opposite side. The body naturally follows this movement. In this way, the whole body, nose, chest, abdomen, neck, and extremities rest one side at a time.

Bet you didn't know your schnozz was so smart. Our motto is "One nostril stuffed is better than two."

Submitted by Richard Aaron of Toronto, Ontario.

Why Do New Fathers Pass Out Cigars?

"What this country needs is a good five-cent cigar" might have been first uttered in the early twentieth century, but in the late seventeenth and early eighteenth centuries, cigars cost much more than five cents. According to Norman Sharp, of the Cigar Association of America, "cigars were so rare and treasured that they were sometimes used as currency."

Two hundred years ago, a baby boy was considered a valuable commodity. He would work the fields all day and produce money for the father, whereas a baby girl was perceived as a financial drain. At first, precious cigars were handed out as a symbol of celebration only when a boy was born.

By the twentieth century, some feminist dads found it in their hearts to pass the stogies around even when, drat, a girl was born. Now the ritual remains a primitive but relatively less costly act of male bonding—a tribute to male fertility while the poor mother recovers alone in her hospital room.

Submitted by Scott P. Frederick of Wilmington, North Carolina. Thanks also to Mike Bartnik of Omaha, Nebraska; and Dan and Patty Poser of San Luis Obispo, California.

DAVID FELDMAN

What Are Dimples? And Why Do Only Some People Have Them?

Dimples are a generic name for indentations of the skin. Dimples are produced when muscle fibers are attached to the deep surface of the skin, such as in the cheek or chin, or where the skin is attached to bones by fibrous bands, such as the elbow, shoulder, and back.

Dimples are most likely to appear where the skin is most tightly attached to the underlying bone. Anatomist Dr. William Jollie, of the Medical College of Virginia, indicates that "dimples probably are due to some developmental fault in the connective tissue that binds skin to bone."

So all this time we've envied those with dimples but didn't realize that they were exhibiting an anatomical flaw! And the tendency toward dimples seems to be hereditary. You have your father to blame, Michael Douglas.

Submitted by Donna Lamb of Stafford, Texas.

Why Do Bath Towels Smell Bad After a Few Days When They Are Presumably Touching Only Clean Skin?

Most towels are made of 100% cotton. While it's true that after a shower you have eliminated most of the germs and dirt from your skin, the process of rubbing a towel against the body rubs off dead skin that sticks to the moist towel. Towels become an ideal nesting place for the mildew endemic to humid bathrooms.

Most people flip a fan on or open the windows when showering but then turn off the fan or close the windows when they dry themselves. Jean Lang, director of Marketing at Fieldcrest, says it is much more important to promote circulation *after* the shower. Without dispersing the moisture, the bathroom becomes like a terrarium. The same type of mildew that afflicts plastic shower curtains attacks towels, especially if the towels have never dried completely from their last use.

We remember our windowless high school locker room with little nostalgia. The lack of ventilation and circulation led to mildew and smelly towels. We would have gladly endured the smell of garbage for the odious aroma of schoolmates' moist towels.

Submitted by Merry Phillips of Menlo Park, California. Thanks also to Paul Funn Dunn of Decatur, Illinois.

How Do Stamp Pads Keep Moist When They Are Constantly Exposed to the Drying Influence of Air?

The ink used in stamp pads has a glycol and water base, which forms a mixture that actually absorbs moisture from the air. On a humid day, this hygroscopic effect allows the stamp or stamp pad to replenish any moisture lost on dry days.

Submitted by Russ Tremayne of Auburn, Washington.

DAVID FELDMAN

Why Tupperware is sold in private!!!

Why Are Tupperware® Brand Products Sold Only at Parties? Couldn't They Make More Money by Selling the Stuff in Stores Too?

Until Earl Tupper came along, most housewares were made of glass, ceramics, wood, or metal—traditional, dependable materials. In 1945, Tupper established Tupper Plastics and tried to market his containers in retail stores.

Tupper's products bombed. Consumers feared that plastic material would prove flimsy, and they didn't understand or believe that Tupper's innovative airtight seal would keep foods fresh. Two salespeople with experience demonstrating Stanley Home Products on the party plan saw Tupper's products and convinced him that sales would mushroom if his plasticware were demonstrated. Early tests were highly successful. In 1951, Tupperware Home Parties was incorporated, and all Tupper products were removed from store shelves.

Now that the public has learned that Tupperware plastic is durable and effective, why doesn't Tupperware compete with less established brand names in K-Mart's and Macy's? Tupperware is convinced that the party approach has unique advantages. Lawrie Pitcher Platt, Tupperware's director of Public Relations and Community Affairs, explains:

> Tupperware brand products continue to be sold on the party plan because each dealer is like a teacher. He or she demonstrates the many subtle features designed into the pieces shown and discusses product care and the full lifetime warranty. Tupperware brand products are a lifetime purchase, unlike many products manufactured today, and it is management's belief that learning about use and care enhances the value to the customer.

Translation: The Tupperware dealer justifies the higher cost of its product.

As of 1989, Tupperware has 89,000 independent dealers in the United States alone and 325,000 in forty-two countries worldwide. With such a solid sales force base, Tupperware would jeopardize the revenue of its dealers by selling Tupperware brand products on a retail basis again. Why risk a retail rollout when Tupperware already has a dedicated sales force devoted solely to its product? Avon and Fuller Brush have experienced problems with direct sales of late, but Tupperware's success may be partly attributable to its party concept, in which the "sponsor" gets rewarded with free merchandise for throwing the party. And unlike many direct sellers, Tupperware doesn't necessarily invade customers' homes. About 25% of all Tupperware parties in the United States are now held outside the home.

Submitted by Charles Kluepfel of Bloomfield, New Jersey.

Why Do Monkeys in the Zoo Pick Through Their Hair All the Time? Why Do They Pick Through One Another's Hair?

In the wild, primates pick at their own hair frequently but for short periods of time. Usually, they are trying to rid themselves of parasitic insects, insect webs, or remnants of food.

DAVID FELDMAN

Monkeys in captivity are much less likely to be riddled with parasites, but may be afflicted with another skin problem. Monkeys exude salt from the pores of their skin. The salt lands on loose bits of skin, and monkeys will often pick through their hair trying to shed the salty flakes.

A monkey, unlike a human, has no difficulty in scratching its back (or any other part of its body, for that matter). Most animal behaviorists assume that apes—be they gibbons or chimpanzees—search through one another's hair for purely social reasons. One psychologist, H. H. Reynolds, noted that chimpanzees are not altruistic or naturally cooperative: "Grooming behavior appears to be one of the most cooperative ventures in which chimpanzees engage."

Perhaps mutual grooming in monkeys is akin to the human handshake, whose original purpose was to signal that a potential weapon, the outstretched hand, would not turn into a clenched fist.

Why Is Cheddar Cheese Orange?

Unless they've been breeding some pretty strange cows in Wisconsin, we would expect cows to produce white milk. All the folks in the dairy industry assured us that they haven't bred a mutant race of cows just to produce orangeish cheddar cheese.

Cheddar cheese is artificially colored with natural ingredients, most commonly annatto, a seed obtained from the tropical annatto tree, found in Central America. Kraft, the largest seller of cheese in the United States, uses a combination of annatto and oleoresin paprika, an oil extraction of the spice paprika, to color its cheddar cheese. Depending upon the natural color of the milk and the amount of annatto added, cheese can be turned into a bright orange color or a more natural-looking yellow shade.

The only reason why cheesemakers color their product is because consumers seem to prefer it. Regional tastes differ, though. Some areas of the eastern United States prefer white cheese, while most of the rest of the country favors yellow. Kraft even makes white "American Singles," although the artificially colored yellow slices far outsell them.

Submitted by Christoper S. von Guggenberg of Alexandria, Virginia.

What Is the Circle Adjacent to the Batter's Box on Baseball Fields?

This area is known as the fungo circle. Coaches stand in the fungo circle during pregame practice and hit balls to infielders and, more frequently, outfielders.

Why confine the coach to stand in one small area? So he won't wear out the grass on the field!

Submitted by Terrell K. Holmes of New York, New York. Thanks also to Ronald C. Semone of Washington, D.C.

What Exactly Is One Hour Martinizing?

Countless millions have passed dry-cleaning stores with the words ONE HOUR MARTINIZING emblazoned on the sign and wondered: What the heck is "Martinizing"? Can it really be done in one hour? Is it painful, and if so, can an anesthetic be administered?

Don't worry. Be happy. Martinizing is a service mark of Martin Franchises, Inc., the largest chain of franchised drycleaning establishments in the United States. Martinizing was first registered with the U.S. Patent Trademark Office in 1950 by

DAVID FELDMAN

the Martin Equipment Corporation, a manufacturer of dry-cleaning machines.

The equipment business and trademarks were later sold to the American Laundry Machinery Company of Cincinnati, Ohio, also a manufacturer of cleaning equipment. Although Martinizing was once part of the sales division of the American Laundry Machinery Company, it has spun off into a separate entity, still located in Cincinnati.

Today if an aspiring dry cleaner wants the know-how and name recognition that a franchise can provide, he or she will likely choose Martin, since it is the best-known name in the dry-cleaning field, and start-up costs are relatively low.

What's special about One Hour Martinizing? As far as we can tell, nothing. They use the same chemicals, solvents, and cleaning methods as other dry cleaners, and can "Martinize" in one hour, just as most dry cleaners can handle a job in one hour.

The folks are relying on the notion that if you patronize another establishment, you can say your clothes have been dry cleaned but you can't brag that they've been Martinized.

Submitted by Dominic Orlando of Arlington, Texas. Thanks also to Peter B. Child of Seattle, Washington.

What Flavor Is Bubble Gum Supposed to Be? Why Is Bubble Gum Usually Pink?

Although in *Imponderables* we managed to ascertain the main flavors in Juicy Fruit gum, we have failed miserably at obtaining the constituents in bubble gum. Perhaps we are losing our powers of persuasion. The best we have been able to wangle from our sources is that "regular" pink bubble gum is a mixture of several natural and artificial fruit flavors.

We thought that the pink color of bubble gum would provide clues to the identity of the flavors, but we were disappointed again. Bubble gum was invented in 1928 by a lone entrepreneur, Walter Diemer, who was an accountant from Philadelphia. From the very beginning, Diemer artificially colored his gum pink. Why? "Because it was the only coloring I had handy at the time!" So much for the sanctity of pink bubble gum.

Now, of course, with Bubble Yum coming in flavors like Bananaberry Split and Checkermint, pink bubble gum looks old hat. But not quite yet. Good old pink bubble gum is still the best seller by far.

Submitted by John Geesy of Phoenix, Arizona.

Why Don't Traffic Signal Light Bulbs Ever Seem to Burn Out? Can We Buy Them?

To answer the second part of the Imponderable first: sure, you can buy the same bulbs that light our traffic signals. But you probably wouldn't want to buy them.

Yes, the bulbs found in traffic lights do last much longer than standard household bulbs. The traffic light bulbs are rated at eight thousand hours, compared to the standard one thousand hours. Incandescent lights can be manufactured to last any length of time. However, the longer life a bulb has, the less efficiently it burns. According to General Electric's J. Robert Moody:

> The incandescent light is like a candle. If you burn it dimly, the candle will last a long time. If you burn the candle on both ends, you get a lot of light but short life. The traffic signal light must use 100 watts to get 1,000 lumens [units of light]. To obtain the same 1,000 lumens a household lamp needs only 60 watts. At an electric rate of $0.10/Kwh, the electric cost for 100 watts is $10.00 per 1,000 hours. For the 60 watts the electric cost is $6.00 per 1,000 hours. Thus, the consumer saves $4.00 per 1,000 burning hours [or 40%] by using a household light bulb rather than a traffic signal light bulb.

Traffic signal bulbs are also specially constructed and are filled with krypton gas rather than the less expensive argon gas used in standard bulbs. Municipalities obviously feel the added expense of the special bulbs is more than offset by the cost of labor for replacing burned-out bulbs and the fewer dangerous situations created by malfunctioning traffic signals.

WHEN DO FISH SLEEP?

We're as lazy as the next guys, but even we figure it is worth changing bulbs to save nearly 50% on our lighting needs. Now if we could get a flashing red light, that might be worth it . . .

Submitted by Michael B. Labdon of Paramount, California.

Why Does Mickey Mouse Have Four Fingers?

Or more properly, why does Mickey Mouse have three fingers and one thumb on each hand? In fact, why is virtually every cartoon animal beset with two missing digits?

Conversations with many cartoonists, animators, and Disney employees confirm what we were at first skeptical about. Mickey Mouse has four fingers because it is convenient for the artists and animators who have drawn him. In the early cartoons, each frame was hand-drawn by an animator—painstaking and tedious work. No part of the human anatomy is harder to draw than a hand, and it is particularly difficult to draw distinct fingers without making the whole hand look disproportionately large.

The artists who drew Mickey were more than happy to go along with any conceit that saved them some work. So in Disney and most other cartoons, the animals sport a thumb and three fingers, while humans, such as Snow White and Cinderella, are spared the amputation.

And before anyone asks—no, we don't know for sure *which* of Mickey's fingers got lopped off for the sake of convenience. Since the three nonthumbs on each hand are symmetrical, we'd like to think it was the pinkie that was sacrificed.

Submitted by Elizabeth Frenchman of Brooklyn, New York.
Thanks also to R. Gonzales of Whittier, California.

Why Don't Migrating Birds Get Jet Lag? Or Do They?

No, birds don't seem to suffer from jet lag. But then again they don't suffer from airport delays, crowded seating, inedible airline food, or lost luggage either.

Human jet lag seems to be bound inextricably to passing rapidly through time zones. Birds usually migrate from north to south, often not encountering any time change. Veterinarian Robert B. Altman speculates that if you put a bird on an airplane going east to west, it might feel jet lag.

But birds, unlike humans, don't try to fly from New York to Australia in one day. Some migrations can take weeks. Birds don't stretch their physical limits unless they have to (such as when flying over a large body of water). If they are tired, birds stop flying and go to sleep, while their human counterparts on the airplane choose between being kept awake by a screaming baby or the one movie they have assiduously avoided seeing in its theater or cable presentations.

Humans are particularly susceptible to jet lag when they travel at night. As a rule, migration doesn't upset birds' natural sleeping patterns. They sleep when it is dark and awaken when it is light. On airplanes, humans fall asleep only immediately preceding the meal service or the captain's latest announcement of the natural wonders on the ground.

Of course, migration isn't without some perils of its own. The National Audubon Society sent *Imponderables* an article detailing the migration habits of shore birds along the Delaware Bay. Many of these shore birds travel from their breeding ground in the Arctic to the southern tip of South America. The round trip can be in excess of fifteen thousand miles.

When the birds land in warmer climes, they engage in a feeding frenzy not unlike a season-long Thanksgiving dinner. The birds found in the Delaware Bay, who had often flown more than five thousand miles with little rest, often doubled their body weight in two weeks. An official of the New Jersey Division of Fish, Game and Wildlife is quoted as saying that the birds "get so fat they can hardly even fly." *New York Times* reporter Erik Eckholm describes these fatted birds as bouncing along "like an overloaded airplane when trying to take off."

Submitted by Chris Whelan of Lisle, Illinois.

Why Do Some Hard-Boiled Egg Yolks Turn Gray or Green When Soft-Boiled Eggs Don't Discolor?

The discoloring is caused by iron and sulphur compounds that accumulate when eggs are overcooked. Although gray egg yolks lack eye appeal, the iron and sulfur don't affect the taste or nutritional value of the eggs.

Probably the most common way of overcooking eggs is to leave the eggs in hot water after cooking. The American Egg Board recommends that after eggs are cooked either cold water should be run over them or they should be put in ice water until completely cooled. Cooling eggs in this manner will not only avoid overcooking but will also make the shells much easier to peel.

DAVID FELDMAN

Why Are Tennis Balls Fuzzy?

The core of a tennis ball is made out of a compound consisting of rubber, synthetic materials, and about ten chemicals. The compound is extruded into a barrel-shaped pellet that is then formed into two half shells.

The edges of the two half shells are coated with a latex adhesive and then put together and cured in a double-chambered press under strictly controlled temperature and air-pressure conditions. The inner chamber is pressurized to thirteen psi (pounds per square inch), so that the air is trapped inside and the two halves are fused together at the same pressure.

Once the two halves have been pressed together to form one sphere, the surface of the core is roughened so that the fuzz will stick better. The core is then dipped into a cement compound and oven-dried to prepare for the cover application.

The fuzzy material is felt, a combination of wool, nylon, and Dacron woven together into rolls. The felt is cut into a figure-eight shape (one circular piece of felt wouldn't fit as snugly on a ball), and the edges of the felt are coated with a seam adhesive. The cores and edges of the two felt strips are mated, the felt is bonded to the core, and the seam adhesive is cured, securing all the materials and for the first time yielding a sphere that looks like a tennis ball.

After the balls are cured, they are steamed in a large tumbler and fluffed in order to raise the nap on the felt, giving the balls their fuzzy appearance. Different manufacturers fluff their balls to varying degrees. The balls are then sealed in airtight cans pressurized at twelve to fifteen psi, with the goal of keeping the balls at ten to twelve psi.

The single most expensive ingredient in a tennis ball is the felt. Many other sports do quite well with unfuzzy rubber balls. In the earliest days of tennis, balls had a leather cover, and were stuffed with all sorts of things, including human hair. So why do tennis ball manufacturers bother with the fuzz?

Before the felt is added, a tennis ball has a hard, sleek surface, not unlike a baseball's. One of the main purposes of the fuzz is to slow the ball down. The United States Tennis Association maintains strict rules concerning the bound of tennis balls. One regulation stipulates, "The ball shall have a bound of more than 53 inches and less than 58 inches when dropped 100 inches upon a concrete base." The fluffier the felt, the more wind resistance it offers, decreasing not only the bound but the speed of the ball. If the felt were too tightly compacted, the ball would have a tendency to skip on the court.

A second important reason for fuzzy tennis balls is that the fluffy nap contributes to increased racket control. Every time a tennis ball hits a racket the strings momentarily grip the ball, and the ball compresses. With a harder, sleeker surface, the ball would have a tendency to skip off the racket and minimize the skill of the player.

A third contribution of fuzz is the least important to a good player but important to us refugees from hardball sports like racquetball and squash. When you get hit hard by a fuzzy tennis ball, you may want to cry, but you don't feel like you're going to die.

Submitted by Dorio Barbieri of Mountain View, California.

DAVID FELDMAN

What Causes Floaters, or Spots, in the Eyes?

The innermost part of the eye is a large cavity filled with a jelly-like fluid known as vitreous humor. Floaters are small flecks of protein, pigment, or embryonic remnants (trapped in the cavity during the formation of the eye) that suspend in the vitreous humor.

The small specks appear to be in front of the eye because the semitransparent floaters are visible only when they fall within the line of sight. Most people might have specks trapped in the vitreous humor from time to time but not notice them. Eyes have a way of adjusting to imperfections, as any eyeglass wearer with dirty lenses could tell you. Floaters are most likely to be noticed when one is looking at a plain background, such as a blackboard, a bare wall, or the sky.

What should one do about floaters? An occasional spot is usually harmless, although sometimes floaters can be precursors of retinal damage. Most often, a home remedy will keep floaters from bothering you. The American Academy of Ophthalmology suggests:

> if a floater appears directly in your line of vision, the best thing to do is to move your eye around, which will cause the inside fluid to swirl and allow the floater to move out of the way. We are most accustomed to moving our eyes back and forth, but looking up and down will cause different currents within the eye and may be more effective in getting the floaters out of the way.

Although you may be aware of their presence, it is often surprisingly difficult to isolate floaters in your line of vision. Because the floaters are actually within the eye, they move as your eyes move and seem to dart away whenever you try to focus on looking at them directly.

Submitted by Gail Lee of Los Angeles, California.

Does It Ever Really Get Too Cold to Snow?

Having withstood a few snowy midwestern winters in our time, we're not sure we would want to test this hypothesis personally. Luckily, meteorologists have.

No, it never gets too cold to snow, but at extremely low temperatures the amount of snow accumulation on the ground is likely to be much lower than at 25 degrees Fahrenheit. According to Raymond E. Falconer, of the Atmospheric Sciences Research Center, SUNY at Albany, there is so little water vapor available at subzero temperatures that snow takes the shape of tiny ice crystals, which have little volume and do not form deep piles. But at warmer temperatures more water vapor is available, "so the crystals grow larger and form snowflakes, which are an agglomerate of ice crystals." The warmer the temperature is, the larger the snowflakes become.

What determines the size of the initial snow crystals? It depends upon the distribution of temperature and moisture from the ground up to the cloud base. If snow forming at a high level drops into much drier air below, the result may be no accumulation whatsoever. In the condition known as "virga," streaks of ice particles fall from the base of a cloud but evaporate completely before hitting the ground.

Submitted by Ronald C. Semone of Washington, D.C.

Why Do Dogs Have Black Lips?

You would prefer mauve, perhaps? Obviously dogs' lips have to be some color, and black makes more sense than most.

According to veterinarian Dr. Peter Ihrke, pigmentation helps protect animals against solar radiation damage. Because

DAVID FELDMAN

dogs don't have as much hair around their mouths as on most parts of their bodies, pigmentation plays a particularly important role in shielding dogs against the ravages of the sun.

According to Dr. Kathleen J. Kovacs, of the American Veterinary Medical Association, the gene for black pigment is dominant over the genes for all other pigments, so the presence of black lips is attributable to hereditary factors. If two purebred dogs with black lips breed, one can predict with confidence that their puppies will have black lips too.

Not all dogs have black lips, though. Some breeds have nonpigmented lips and oral cavities. James D. Conroy, a veterinary pathologist affiliated with Mississippi State University, told *Imponderables* that some dogs have a piebald pattern of nonpigmented areas alternating with pigmented areas. The only breed with an unusual lip color is the Chow Chow, which has a blue color. Conroy says that "the blue appearance of the lips and oral cavity is related to the depth of the pigment cells within the oral tissue."

Submitted by Michael Barson of Brooklyn, New York.

If Church and State Are Supposed to be Separated in the United States, Why Do We Swear On Bibles in Courts? What Happens if a Witness Doesn't Accept the Validity of the Bible?

The ritual of taking an oath with the right hand raised while placing the left hand on a holy object goes back to ancient times. Michael De L. Landon, secretary of the American Society for Legal History, sent us a picture of the Bayeux Tapestry, which depicts King Harold of England, who reigned from 1035–1040, taking an oath with both hands on a sacred object.

In the Middle Ages, before printed Bibles were commonly

available, Christians placed the left hand on a relic of a saint or some other sacred object and raised the right hand while taking an oath. Professor De L. Landon comments:

> the right hand raised and open, palm outward, is an internationally recognized gesture implying peace, honesty, and good intentions. In taking an oath, there is also probably the indication that one is pointing to heaven and calling upon God (or the gods) to be one's witness that one is sincere and telling the truth.

The English adopted the practice of having witnesses swear an oath on the bible before testifying. American law was based originally on an English common law that stipulated that only witnesses who believed in a Supreme Being could testify at a trial. The framers of the common law assumed that only the fear of an eternal punishment would ensure the honesty of the witness. Lord Coke, the leading English jurist of the early seventeenth century, went further and argued that nonconformists as well as atheists were *petui inimici* ("eternal enemies") and should be barred from testifying. Coke's position was adopted by the English for almost two hundred years, and although it became impossible to enforce the doctrine, Parliament did not actually remove the statute until 1869.

Most courtrooms have stopped using Bibles to swear in witnesses, for the ritual was always a ceremonial demonstration of good faith rather than a legally mandated procedure. Most courts traditionally have used King James Bibles, but have allowed Jews or Catholics to substitute versions that were acceptable to their faith.

The United States adopted the rule disqualifying disbelievers in the Federal Judiciary Act of 1789, which provided that no one could testify "who did not believe that there is a God who rewards truth and avenges falsehood." In 1906 Congress passed an amendment to allow states to determine their own rules for their own courts, although most states had already passed statutes voiding the disbeliever clause. Even today, a few states have not struck down the disbeliever clause; theoretically, an atheist could be barred from testifying in a trial in those states.

DAVID FELDMAN

In his book *Church, State and Freedom*, Leo Pfeffer notes broader constitutional provisions ensure that no civil rights may be denied because of religious beliefs. Still, the issue hasn't been addressed squarely by the Supreme Court, and Pfeffer documents a scary application of how the nonbeliever clause has been applied in the past:

> in 1900 the Court upheld a Federal statute that required that the testimony of Chinese, in certain cases, be corroborated by that of white men, because of the 'loose notions entertained by [Chinese] witnesses of the obligation of an oath.' It would seem clear that if a defendant in a criminal case or a party in a civil case could not take the stand in his own behalf because of his religious beliefs or disbeliefs, he would be deprived of his liberty or property without due process of law and would be denied the equal protection of the laws in violation of the Fourteenth Amendment. Moreover ... the 'free exercise' clause of the First Amendment protects religious disbelief as well as belief ...

As might be expected, both the ACLU and Madalyn Murray-O'Hair's Society of Separationists have been in the forefront of litigation attempting to eliminate swearing on Bibles (as well as eliminating other elements of religion in the courts). The path of least resistance, for most jurisdictions, has been to abandon the use of Bibles.

One need not be philosophically opposed to the use of Bibles in the courtroom to note that a Bible has never been a guarantee of truthful testimony; perjurers have been swearing on Bibles for a long, long time.

Why Do Females Tend to Throw "Like a Girl"?

Not only do girls (and later, women) tend not to be able to throw balls as far as boys, but their form is noticeably different. If you ask the average boy to throw a baseball as far as he can, he will lift his elbow and wind his arm far back. A girl will tend to keep her elbow static and push forward with her hand in a motion not unlike that of a shot putter.

Why the difference? Our correspondent mentions that he has heard theories that females have an extra bone that prevents them from throwing "like a boy." Or is it that they are missing one bone?

We talked to some physiologists (who assured us that boys and girls have all the same relevant bones) and to some specialists in exercise physiology who have studied the underperformance of girls in throwing.

In their textbook, *Training for Sport and Activity: The Physiological Basis of the Conditioning Process*, Jack H. Wilmore

DAVID FELDMAN

and David L. Costill cite quite a few studies that indicate that up until the ages of ten to twelve, boys and girls have remarkably similar scores in motor skills and athletic ability. In almost every test, boys barely beat the girls. But at the onset of puberty, the male becomes much stronger, possesses greater muscular and cardiovascular endurance, and outperforms girls in virtually all motor skills.

In only one athletic test do the boys far exceed the girls before and after puberty: the softball throw. From the ages of five to sixteen, the average boy can throw a softball about twice as far as a girl.

Wilmore and Costill cite a fascinating study that attempted to explain this phenomenon. Two hundred males and females from ages three to twenty threw softballs for science. The result: males beat females two to one when throwing with their dominant hand, but females threw almost as far as males with their nondominant hand. Up until the ages of ten to twelve, girls threw just as far with their nondominant hand as boys did.

The conclusion of Wilmore and Costill is inescapable:

> Major differences at all ages were the results for the dominant arm . . . the softball throw for distance using the dominant arm appears to be biased by the previous experience and practice of the males. When the influence of experience and practice was removed by using the nondominant arm, this motor skill task was identical to each of the others.

All the evidence suggests that girls can be taught, or learn through experience, how to throw "like a boy." Exercise physiologist Ralph Wickstrom believes most children go through several developmental stages of throwing. Boys simply continue growing in sophistication, while girls are not encouraged to throw softballs or baseballs and stop in the learning curve. As an example, Wickstrom notes that most right-handed girls throw with their right foot forward. Simply shifting their left foot forward would increase their throwing distance.

When forced to throw with their nondominant hand, most boys throw "like a girl." The loss in distance is accountable not only to lesser muscular development in the nondominant side, but to a breakdown in form caused by a lack of practice.

Submitted by Tony Alessandrini of Brooklyn, New York.

Given that the ZIP Code Defines the City and State, Why Do We Have to Include Both on Envelopes? Or Do We?

Jack Belck, the true zealot who posed this Imponderable, gave as his return address his full name, a street number, and 48858, with a note: "The above address is guaranteed to work."

Evidently it did. He received a letter we wrote to him in that lovely town, 48858.

But the question is a good one, so we asked our friends at the USPS to respond.

And they were a tad cranky.

Yes, they will deliver letters addressed by the Belcks of the world, but they aren't too happy about it for a couple of reasons. First of all, many people inadvertently transpose digits of the ZIP code. The city and state names then serves as a cross check. Without the city and state names, the letter would be returned automatically to the sender. Even if it is delayed, the postal service will reroute a letter with an incorrect ZIP code.

Secondly, Mr. Belck isn't quite right about one of his premises. In rural areas, more than one municipality might share the same zip code. City names can thus be of assistance to the local post office in sorting and delivering the mail.

Submitted by Jack Belck of 48858.

DAVID FELDMAN

Why Do Telephone Cords Spontaneously Twist Up? What Can One Do About this Dreaded Affliction?

Spontaneously twist up, you say? You mean you sit on your sofa watching TV and suddenly the telephone cord starts winding like a snake?

After considerable research into the matter, we must conclude that telephone cords do not twist up spontaneously. You've been turning around the headset, Alan. We're not accusing you of doing this intentionally, mind you. As far as we know, twisting a headset is not even a misdemeanor in any state or locality. But don't try to blame your indiscretions on the laws of nature. Cords don't cause twisted cords—people do.

Now that we've chastised you, we'll offer the obvious, simple yet elegant solution. Remove the plug that connects the headset to the body of the phone. Hold the cord by the plug side and let the headset fall down (without hitting the floor, please). The cord will "spontaneously" untwist.

For those having similar problems with twisted lines connecting their phones to the modular jacks in the wall, simply unplug the line from the phone. If the line is sufficiently coiled, it will untwist like an untethered garden hose.

Submitted by Alan B. Heppel of West Hollywood, California.

Why Do Golf Balls Have Dimples?

Because dimples are cute?

No. We should have known better than to think that golfers, who freely wear orange pants in public, would worry about cosmetic appearances.

Golf balls have dimples because in 1908 a man named Taylor patented this cover design. Dimples provide greater aerodynamic lift and consistency of flight than a smooth ball. Jacque Hetric, director of Public Relations at Spalding, notes that the dimple pattern, regardless of where the ball is hit, provides a consistent rotation of the ball after it is struck.

Janet Seagle, librarian and museum curator of the United States Golf Association, says that other types of patterned covers were also used at one time. One was called a "mesh," another the "bramble." Although all three were once commercially available, "the superiority of the dimpled cover in flight made it the dominant cover design."

Although golfers love to feign that they are interested in accuracy, they lust after power: Dimpled golf balls travel farther as well as straighter than smooth balls. So those cute little dimples will stay in place until somebody builds a better mousetrap.

Submitted by Kathy Cripe of South Bend, Indiana.

Why Don't Crickets Get Chapped Legs from Rubbing Their Legs Together? If Crickets' Legs Are Naturally Lubricated, How Do They Make that Sound?

If we rubbed our legs together for five minutes as vigorously as crickets do all the time, our legs would turn beet red and we would hobble into the bathroom searching for the talcum powder. How do crickets survive?

Quite well, it turns out. For it turns out that we can't believe everything we learned in school. Crickets don't chirp by rubbing their legs together. Entomologist Clifford Dennis explains:

> Crickets do not produce chirps by rubbing their legs together. They have on each front wing a sharp edge, the scraper, and a file-like ridge, the file. They chirp by elevating the front wings and moving them so that the scraper of one wing rubs on the file of the other wing, giving a pulse, the chirp, generally on the closing stroke.

On a big male cricket, the scraper and the file can often be seen by the naked eye. You can take the wings of a cricket in your fingers and make the chirp sound yourself.

No thanks. We'll take your word for it on faith.

Submitted by Sandra Baxter of Ada, Oklahoma.

Why Is a Navy Captain a Much Higher Rank than an Army Captain? Has This Always Been So?

When one looks at the ranks of the officers of the four branches of the American military, one is struck by how the Army, Air Force, and Marine Corps use the identical ranks, while the Navy uses different names for the equivalents. But there is one striking disparity: the Navy elevates the rank of captain.

Army, Air Force, Marine Corps	Navy
Warrant Officer	Warrant Officer
Chief Warrant Officer	Chief Warrant Officer
Second Lieutenant	Ensign
First Lieutenant	Lieutenant Junior Grade
CAPTAIN	Lieutenant
Major	Lieutenant Commander
Lieutenant Colonel	Commander
Colonel	CAPTAIN
Brigadier General	Commodore
Major General	Rear Admiral
Lieutenant General	Vice Admiral
General	Admiral
General of the Army or General of the Air Force	Fleet Admiral

The word "captain" comes from the Latin word *caput*, meaning "head." In the tenth century, captains led groups of Italian foot soldiers. By the eleventh century, British captains commanded

DAVID FELDMAN

warships. So the European tradition has been to name the head of a military unit of any size, on land or sea, a captain.

Our elevation of the English captain stems from English naval practice. In the eleventh century, British captains were not the heads of ships *per se*. Although captains were in charge of leading soldiers in combat aboard ship, the actual responsibility for the navigation and maintenance of ships fell upon the ranks of master. By the fifteenth century, captains bristled at deferring to the masters they outranked, and captains began to assume the responsibility for the ships heretofore claimed by masters. By 1747 any commander of a ship was officially given the rank of captain.

Meanwhile, on land most European countries named the commander of a company—of any size—captain. By the sixteenth century, military strategists felt that one hundred to two hundred men were the maximum size for a land unit in battle to be effectively led by one person. That leader was known as a captain.

From the inception of the United States military we borrowed from the European tradition. A captain was a company commander and indeed is so today. In the Air Force, a captain commands a squadron, the airborne equivalent of a company. But the Navy captain, because he has domain over such a big and complicated piece of equipment, has a legitimate claim to a higher rank than his compatriots in the other branches. As Dr. Regis A. Courtemanche, of the Scipio Society of Naval and Military History, put it,

> Navy captain isn't only a rank. The senior officer of a ship is always called "Captain" even though his rank may only be lieutenant. So a naval captain may have more responsibility than a military captain who usually commands only a small detachment in battle.

In 1862, the Navy realized that it was no longer practical to make captain its highest rank. They needed a way not only to differentiate among commanders of variously sized and equipped vessels but to reward those who were supervising the captains of warships. For this reason, the Navy split the rank of captain into three different categories. The commodore (and later, the rear admiral) became the highest grade, the commander the lowest, and the captain, once ruler of the seas, stuck in the middle of the ranks.

Submitted by Barrie Creedon of Philadelphia, Pennsylvania.

Why Do Astronomers Look at the Sky Upside Down and Reversed? Wouldn't It Be Possible to Rearrange the Mirrors on Telescopes?

Merry Wooten, of the Astronomical League, informs us that most early telescopes didn't yield upside-down images. Galileo's original spyglass used a negative lens as an eyepiece, just as cheap field glasses made with plastic lenses do now. So why do unsophisticated binoculars yield the "proper" image and expensive astronomical telescopes render an "incorrect" one?

Astronomy editor Jeff Kanipe explains:

> The curved light-gathering lens of a telescope bends, or refracts, the light to focus so that light rays that pass through the top of the lens are bent toward the bottom and rays that pass through the bottom of the lens are bent toward the top. The image thus forms upside down and reversed at the focal point, where an eyepiece enlarges the inverted and reversed image.

Alan MacRobert, of *Sky & Telescope* magazine, adds that some telescopes turn the image upside down, and others also mirror-reverse it: "An upside-down 'correct' image can be viewed correctly just by inverting your head. But a mirror image does not become correct no matter how you may twist and turn to look at it."

O.K. Fine. We could understand why astronomers live with inverted and upside-down images if they had to, but they don't.

DAVID FELDMAN

Terrestrial telescopes do rearrange their image. Merry Wooten says that terrestrial telescopes can correct their image by using porro prisms, roof prisms, or most frequently, an erector lens assembly, which is placed in front of the eyepiece to create an erect image.

Why don't astronomical telescopes use erector lenses? For the answer, we return to Jeff Kanipe:

> Most astronomical objects are very faint, which is why telescopes with larger apertures are constantly being proposed: Large lenses and mirrors gather more light than small ones. Astronomers need every scrap of light they can get, and it is for this reason that the image orientation of astronomical telescopes are not corrected. Each glass surface the light ray encounters reflects or absorbs about four percent of the total incoming light. Thus if the light ray encounters four glass components, about sixteen percent of the light is lost. This is a significant amount when you're talking about gathering the precious photons of objects that are thousands of times fainter than the human eye can detect. Introducing an erector into the optical system, though it would terrestrially orient the image, would waste light. We can afford to be wasteful when looking at bright objects on the earth but not at distant, faint galaxies in the universe.

And even if the lost light and added expense of erector prisms weren't a factor, every astronomer we contacted was quick to mention an important point: There IS no up or down in outer space.

Submitted by William Debuvitz of Bernardsville, New Jersey.

Why Are the Rolls or Bread Served on Airlines Almost Always Cold While Everything Else on the Tray Is Served at the Appropriate Temperature?

We won't even comment on the *taste* of airline food (this is a family book). But if McDonald's can separate the cold from the hot on a McDLT sandwich, why can't the airlines get their rolls within about 50 degrees of the right temperature?

The answer lies in how airline meals are prepared aloft. The salad, bread, and dessert are placed on trays that are usually refrigerated or packed in ice. Entrees are loaded onto separate baking sheets. When it is time to start the meal service, the flight attendant who prepares the meals simply sticks the trays of entrees into ovens (not, by the way, microwaves).

The rolls are cold because they have been sitting all along with the salad and cake. Most airlines offer customers a choice of entrees. The flight attendant who is serving the meal simply selects the entree from the sheets they were cooked in and places it alongside the rest of the meal. Except for the entree choice, every flier's tray will look identical. Note that although most airlines vary the vegetable according to the entree, the vegetable is always cooked on the same plate as the main course because the entree plate will be the only heated element on the tray.

If the bread and salad taste cold, why doesn't the dessert? Airlines, almost without exception, serve cake for dessert. Michael Marchant, vice president of Ogden Allied Aviation Services and the president of the Inflight Food Service Association, told *Imponderables* that the softness of cake fools us into thinking it is being served at room temperature. The gustatory illusion is maintained because in contrast to the roll's hard crust, which locks in the coldness, the soft frosting of a cake dissipates the cold.

The folks in first class, meanwhile, are munching warm rolls, which have been heated. Certainly it is worth an extra five hundred dollars or so to get heated rolls, isn't it?

Why Do Chickens and Turkeys, Unlike Other Fowl, Have White Meat and Dark Meat?

Other birds that we eat, such as quail, duck, or pigeon, have all dark meat. Chickens and turkeys are among a small group of birds with white flesh on the breasts and wings.

Birds have two types of muscle fibers: red and white. Red muscle fibers contain more myoglobin, a muscle protein with a red pigment. Muscles with a high amount of myoglobin are capable of much longer periods of work and stress than white fibers. Thus, you can guess which birds are likely to have light fibers by studying their feeding and migration patterns.

Most birds have to fly long distances to migrate or to find food, and they need the endurance that myoglobin provides. All birds that appear to have all white flesh actually have some red fibers, and with one exception, all birds that appear to be all dark have white fibers. But the hummingbird, which rarely stops flying, has pectoral muscles consisting entirely of red fibers because the pectoral muscles enable the wings to flap continuously.

The domestic chicken or turkey, on the other hand, lives the life of Riley. Even in their native habitat, according to Dr. Phil Hudspeth, vice president of Quality and Research at Holly Farms, chickens are ground feeders and fly only when nesting. Ordinarily, chickens move around by walking or running, which is why only their legs and thighs are dark. They fly so little that their wings and breasts don't need myoglobin. In fact, the lack of myoglobin in the wing and breast are an anatomical advantage. Janet Hinshaw, of the Wilson Ornithological Society, explains why chicken and turkey musculature is perfectly appropriate:

> They spend most of their time walking. When danger threatens they fly in a burst of speed for a short distance and then land. Thus they need flight muscles which deliver a lot of power quickly but for a short time.

Next time you fork up an extra fifty cents for that order of all-white meat chicken, remember that you are likely paying to eat a bird that racked up fewer trips in the air than you have in an airplane.

Submitted by Margaret Sloane of Chapel Hill, North Carolina. Thanks also to Sara Sickle of Perryopolis, Pennsylvania; and Annalisa Weaver of Davis, California.

Why Haven't Vending Machines Ever Accepted Pennies?

In the second half of the twentieth century, when a child is more likely to think that penny candy is the name of a cartoon character rather than the actual price of a confection, it is hard to believe that in the early days of vending machines the industry would have loved to be able to accept pennies. When a candy bar cost five cents, vendors undoubtedly lost many sales when frustrated kids could produce five pennies but not one nickel. Now, when a candy bar might cost half a dollar, payment in fifty pennies might clog a receptacle. But why didn't vending machines *ever* accept pennies? We spoke to Walter Reed, of the National Automatic Merchandising Association, who told us about the fascinating history of this Imponderable.

The vending machine industry has always been plagued by enterprising criminals who inserted slugs or relatively worthless foreign coins into machines in the time-honored tradition of trying to get something for nothing. In the 1930s, a slug rejector was invented that could differentiate U.S. coinage from Mexican centavos of the same size. The slug rejector worked by determining the metallic content of the coin. Although the slug rejector could easily differentiate between silver or nickel and a slug, it couldn't tell the difference between a worthless token and the copper in a penny. For this reason, vendors hesitated to accept pennies in the machines.

The slug rejectors of today are much more sophisticated, measuring the serration of the coin, its circumference, its thickness, and the presence of any holes. Whereas the 1930s slug rejector was electromagnetic, current rejectors perform tests electronically.

The vending machine industry was instrumental in pushing for the clad-metal coins that were introduced in 1965. Since that year our quarter, for example, which used to be made of silver, now has a center layer of copper surrounded by an outer layer of copper and nickel. The copper-nickel combination reacts to the electronic sensors in vending machine rejectors much like silver. The government also loves the clad coins because the constituent metals are so much cheaper to buy.

Except in gumball machines, the vending machine industry has never accepted pennies, although they once gave pennies away to consumers. In the late 1950s, a cigarette tax was imposed that drove the retail price of cigarettes a few cents above its long-held thirty-five-cent price. Stores simply charged thirty-seven cents, but vending machines couldn't, for they were not equipped to return pennies.

Vendors had to decide whether to keep charging thirty-five cents and absorb the loss of the two cents on every pack, or charge forty cents and risk loss of sales when grocery stores could undercut them by 10%. So they compromised. Vending machines charged forty cents a pack, but pennies were placed in the pack to restore equity to the consumer.

Submitted by Fred T. Beeman of Wailuku, Hawaii.

NOW that Most Products Sold in Vending Machines Sell for Fifty Cents or More, Why Don't Most Vending Machines Accept Half Dollars or Dollar Bills?

The problem with the half dollar is that the public does not carry it in its pocket. Half dollars are too bulky and heavy. Allowing half dollars would necessitate increasing the size of coin slots in the machines.

The American public loves quarters. Unfortunately, studies have shown that people resist putting in more than two coins in vending machines. And two quarters aren't enough to buy even a soft drink anymore.

So isn't the dollar bill acceptor the panacea? The technology exists to accept dollar bills in vending machines, but the same hassles that plague the consumer using dollar-bill changers are also a nightmare for the vendor. Bills must be placed in the proper position to be accepted. Worn or slightly torn bills are rejected routinely even though they are perfectly legal tender. And worst of all, dollar bills can't be counted easily by machine. The labor involved in counting paper money is not insignificant.

The vending machine industry lusts after the resuscitation of the silver dollar. Frustrated by the unpopularity of the Susan

B. Anthony dollar, trade groups are now pushing for a new gold-colored dollar with a portrait of Christopher Columbus on the obverse. The Treasury supports the proposal, for although coins are more expensive to manufacture than bills, they last much longer in circulation. Walter Reed points out that no other industrialized nation has an equivalent of a one dollar bill in paper currency anymore. The Canadians were the last to fall, with the Looney dollar, the same size as the ill-fated Susan B. Anthony, replacing their dollar bill.

Why Is a Blue Ribbon Used to Designate First Prize?

Most sources we contacted give credit to the English for introducing the blue ribbon. In 1348, King Edward III of England established the Order of the Garter, now considered one of the highest orders in the world. Ribbons had traditionally been used as a badge of knighthood. Members of the Order of the Garter were distinguished by wearing their dark blue ribbon on their hip.

A second theory presented by S. G. Yasinitsky, of the Orders and Medals Society of America, was new to us:

> Another version of the blue ribbon as meaning the highest achievement may have originated among British soldiers who practiced abstinence by belonging to the various army abstinence groups, especially in India, in the latter part of the nineteenth century. Their basic badge for the first six years' total abstinence was a medal worn on a blue ribbon. Hence a 'blue ribbon unit' was one which was comprised of all men who were sporting a blue ribbon in their buttonhole to denote their sobriety. 'Blue ribbon panel' and 'blue ribbon selection' followed this, I'm sure.

Yasinitsky and others have speculated that our ribbon color schemes might have had an astronomical basis. Blue, the highest award, represented the sky and the heavens, the highest point possible. Red (second prize) represented the sun, which was high up in the sky. Yellow (third prize) represented the stars, once thought to be lower than the sun. Yasinitsky mentions that runners-up in fairs and festivals are often given green ribbons as consolation prizes. The green color probably represents the lowly grass on the ground.

What Is the "Cottage" in Cottage Cheese?

Food historians speculate that cottage cheese was probably the first cheese. And it was undoubtedly made by accident. Some anonymous nomad was probably carrying milk on a camel in the desert and at the end of the day found lumps rather than liquid. And much to the nomad's surprise, the lumps tasted pretty good.

According to the United Dairy Industry Association, cottage cheese was made in the home all over Europe as far back as the Middle Ages. "It was called 'cottage' because farmers made the cheese in their own cottages to utilize the milk remaining after the cream had been skimmed from it for buttermaking."

Submitted by Mrs. K. E. Kirtley of Eureka, California.

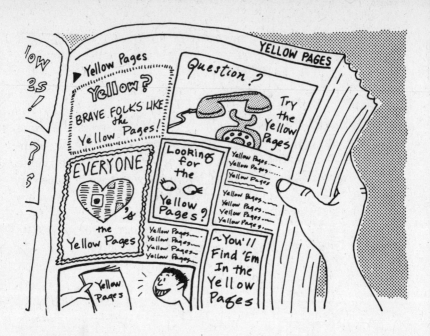

Why Are There So Many Ads for the Yellow Pages in the Yellow Pages?

Yellow Pages publishers are smart enough to realize that if you've got a copy of their directory in your grubby hands, you already are convinced of the efficacy of their medium. So why must they pummel us with promotional ads? Phone companies make profits from their directories by selling advertising space —you would think they'd rather have a local plumber buy a small display ad than toot their own horns.

The simple purpose of the promotional ads is to fill space between paid ads. Kenneth Hudnall, executive director of the National Yellow Pages Agency Association, explains why there is a need for filler:

> Mechanically, the composition of the Yellow Pages is quite involved. For a variety of reasons there will be small bits of space left at the bottom of a column of listings or between display ad-

DAVID FELDMAN

vertisements. Rather than leave this space blank, the publishers will throw in "justifiers" to fill up the space. And what is more natural than to put promotional copy for Yellow Pages in this space?

If all advertisements in the Yellow Pages were the same size, it would be easy for designers to lay out the directory without need for filler. But the ads, whether listings or display, come in all different sizes. A catering company won't want its advertisement stranded alone when all the other caterers in town are listed on the two pages before. Justifiers, then, have been a way to make the life of the designer easier and soothe the complaints of advertisers about the placement of their display ads.

One man, Arnie Nelson, had the kind of brilliant idea that can make fortunes: Why should Yellow Pages publishers "waste" the filler space when they could sell advertising in it? Nelson founded a company called Yellow Spots, Inc., whose purpose is to sell small-space display advertising to companies who traditionally do not advertise at all in the Yellow Pages.

According to Nelson and Yellow Spots executive Gabe Samuels, initially there was some resistance from the regional phone companies to giving Yellow Spots an exclusive right to sell display ads. But Yellow Spots mustered some strong arguments to convince them, the most compelling one economic: it would provide a windfall. According to Yellow Spots, anywhere from 6 to 20% of the Yellow Pages consist of filler. Adding 5 or 10% more to gross revenues through new display ads would be most profitable.

Some of the publishers were also reluctant to introduce a new type of advertising into a medium that had thrived without it for more than a hundred years. Nelson and Samuels argued that the Yellow Pages were actually used more by consumers as an information source, a magazine, rather than as an advertising medium. The editorial matter of the *Yellow Pages Magazine* are the directory listings. Yellow Spots would deliver the advertising, billboard ads without addresses or phone numbers. The ads that Yellow Spots would solicit were designed to promote a product rather than tell consumers where to buy it, thus not alienating Yellow Pages' traditional retail clients.

Yellow Spots' second obstacle was to convince corporations, mostly big, national advertisers, to promote their companies in a medium that had heretofore not been considered. There had never been a category in the Yellow Pages that would allow Coca-Cola to promote the image of its beverage, although local bottlers or distributors might have had their addresses and phone numbers printed.

So how did Yellow Spots attract national advertisers and have the temerity to ask up to $8 million from one potential client? They touted the unique advertising climate that the Yellow Pages presents:

- The circulation of all the Yellow Pages directories in the United States is about 100 million, 10 million more than there are homes in America. The Yellow Pages, of course, is usually used by more than one person.
- 50% of all customer references to the Yellow Pages result in a sale.
- 18% of all adults use the Yellow Pages at least once on any given day (and they average one and one-half uses per day). This is the equivalent of a rating of 18 on TV, emblematic of a successful show.
- Advertisers operate in a nonhostile environment in the Yellow Pages. Whereas the clutter of TV commercials is a bone of contention among viewers, users of the Yellow Pages do not feel oppressed by the number of ads. In a recent survey, 65% of Americans surveyed felt the number of ads in the Yellow Pages were "just about right"; 18% said they wished there were *more* ads; and only 8% complained there were too many ads.
- Yellow Pages are kept in the home all year long and, in many cases, much longer. Magazines—even those passed around within a family—tend to be thrown out within weeks.

Yellow Spots has already signed up Budget car and truck rentals and Sears Discovery card as major accounts, with others soon to follow. Although we admire the ingenuity of Yellow Spots, we're glad that the homely graphics of the promotional

DAVID FELDMAN

fillers won't totally disappear. Even Nelson and Samuels concede that they'll never take over all of the possible remnant space. They will be quite content with about 50 to 60% of it, thank you.

Submitted by Calvin Wong of Chapel Hill, North Carolina.

Why Is Flour Bleached?

Wheat isn't white. Flour is made out of wheat. So why is flour white?

First of all, all of the major flour producers, such as Pillsbury and General Mills, do make unbleached flour, which many breadmakers prefer. But the vast majority of flour sold to consumers is in the form of all-purpose bleached white flour, which is a combination of hard wheat flour (high in protein and best for making breads) and soft wheat flour (lower in protein and the best consistency for cakes and pastries).

Freshly milled white flour has a yellowish tinge, much like unbleached pasta, which consumers reject in favor of a pristine white. Flour processors have two ways to eradicate the yellow from wheat flour. If flour is stored and allowed to age naturally for several months, the yellow disappears as it is exposed to oxygen. But the cost of storing the bulky flour is prohibitive, so commercial flour is bleached artificially with bleaches such as benzoyl peroxide. Artificial bleaching works better than natural aging, which doesn't yield uniformity of color or maturation.

Mature flour produces better baking results and has a longer shelf life. So along with being bleached, all-purpose flour is artificially aged. While benzoyl peroxide merely bleaches flour, other agents such as azodicarbonamide and potassium bromide artificially age the flour as they bleach. The whole process is performed in twenty-four hours, and the bleach eventually decomposes into a harmless residue called benzoic acid when the flour is used.

Is there a down side to the bleaching process? Certain nutrients are lost, which is why all-purpose flour by law is enriched with nutrients. Some nutritionists are not sanguine about the results. The late Adele Davis was particularly rabid about the subject. She felt the machinery that grinds flour overheats it and gives it a precooked taste "comparable to last night's chops reheated." But she was particularly skeptical about the value of enriched flour:

> So-called "enriched" flour is my idea of outright dishonesty; at least 25 nutrients are largely removed during refining, and one-third of the original amount of iron, vitamin B and niacin may be replaced. Such flour is "enriched" just as you would be enriched by someone stealing 25 dollars from you and returning 99 cents.

Flour enrichment was mandated by the federal government in the early 1940s to compensate for the loss of nutrients that are eliminated from white flour. The flour industry contends that Adele Davis and other critics' objections to enrichment overstate the case. Although they concede that the bran and germ of wheat kernels in whole-wheat flour contain more nutrients than white flour, those nutrients lost (e.g., calcium, phosphorus, and potassium) tend to be found in other foods, and few consumers look toward baked goods as a source for these nutrients.

Although health-food advocates tend to belittle the nutritional value of white flour, the flour companies stress that bleaching in itself has never been a health hazard. The alternative to bleached flour, they say, is vastly more expensive flour.

What Is Goofy?

Goofy can't be a dog, claims our correspondent, or else he would look like Pluto, wouldn't he? Goofy is indeed a dog. Chihuahuas don't look like Doberman pinschers, so why should Goofy look like Pluto? Although we must admit that we don't know too

DAVID FELDMAN

many dogs who speak English and walk on two feet.

Pluto appeared several years before Goofy, in a tiny role in a Mickey Mouse short called "Chain Gang." Pluto's original name was Rover, and he was Minnie's dog, not Mickey's. But Mickey soon gained ownership, and Rover was renamed Pluto the Pup. Animator John Canemaker observes that Pluto's lack of speech and doglike walk were used to emphasize that Pluto was Mickey's pet and not his equal.

Goofy, on the other hand, was nobody's pet. His dogginess is indisputable, since his original name was Dippy Dawg. But Dippy had to pay his dues before he reached the summit of Goofyness. Dippy first played small roles in Mickey Mouse shorts in the early 1930s, and it wasn't until he was featured in the syndicated Mickey Mouse newspaper cartoons that he gained prominence in animated shorts.

Although Goofy was as loyal and loving as Pluto, he was not subservient. As his popularity grew, Goofy became a part of "The Gang," with costars Mickey Mouse and Donald Duck in a series of twelve cartoons in the late 1930s and early 1940s. Few remember that Goofy was married (to Mrs. Goofy) and that he was a proud parent (of Goofy, Jr.).

This Imponderable has been thrust at us many times since the release of the movie *Stand By Me*, in which a character muses about this question. How people can accept that a duck can survive being squashed by a refrigerator and then not believe that Goofy can be a dog, we'll never understand.

Submitted by Ashley Hoffar of Cincinnati, Ohio.

HOW Did the Toque Become the Traditional Chef's Hat? Does It Serve Any Functional Purpose?

Most men, in their daily lives, wear neither rags nor haute couture. We don a pair of pants and a shirt—maybe a sports coat or suit and tie if the occasion warrants it. But in the kitchen headware has always been schizophrenic. Cooks wear either ugly but functional hair nets or *toques blanches* ("white caps"), smart-looking caps with tops long enough to camouflage the heads of the entire Conehead family. Isn't there a middle ground? Why can't a chef wear a baseball cap or a derby? Can there possibly be a logical function for the shape of toques?

As early as the Roman and Greek Empires, master chefs were rewarded for their achievements by receiving special headware. For the ancients, laurel-studded caps were the honor.

In France up until the seventeenth century, chefs were awarded different colored caps depending upon their rank. Apprentices wore ordinary skull caps. During the early eighteenth century, Talleyrand's chef required his entire staff to don the

DAVID FELDMAN

toque blanche for sanitary reasons. The toque blanche was designed not only to keep the chef's hair from entering food but to register any stains upon the white background.

But this original cap was flat. The high hat gradually gained popularity not as a fashion statement, not to hide Mohawk hairdos, but to provide some ventilation for the head, as chefs frequently work under extremely hot conditions.

Viennese chef Antonin Careme, not willing to leave well enough alone, decided that the toque blanche needed still more oomph. He put a piece of round cardboard inside his toque to give the cap a stiffer, more dashing appearance. The cardboard has been replaced today by starch.

The toque blanche is no more functional than a hair net, and almost as silly looking. But as Shriners or Mouseketeers can testify, any hat bestowed upon someone as an honor is likely to be worn proudly by the recipient, regardless of how funny it looks.

Submitted by William Lickfield of Hamburg, New York.

When and Where Do Police Dogs Urinate and Defecate?

Our fearless correspondent, Eric Berg, notes that he trains his eyes for police dogs whenever he is in a big city and has yet to see nature call one of our canine protectors. "Have the police bred some sort of Bionic Dog?" Eric wonders.

Natural urges dog police dogs just as often as any Fido or Rover, but the difference is in the training; police dogs are much more disciplined than other dogs, or for that matter, most dog owners. Before the animals go on duty, trainers allow police dogs to run and go to the bathroom (well, not *literally* a bathroom) in the area where they are kept.

Part of the training of police dogs involves teaching the dog to control itself while on patrol and when in front of the public.

The dog is taught to signal when it has to "go," but is trained to keep itself under control in all circumstances.

Gerald S. Arenberg, editor of the official journal of the National Association of Chiefs of Police, alludes to the fact that "the dogs are given walks and care that is generally not seen by the public," the only hint we received that occasionally a dog might relieve itself while on duty.

Let's end this discussion here, before we run out of euphemisms.

Submitted by Eric Berg of Chicago, Illinois.

How Can Hurricanes Destroy Big Buildings But Leave Trees Unscathed?

Think of a hurricane as heavyweight boxer Sonny Liston, a powerful force of nature. A building in the face of Liston's onslaught is like George Foreman, strong but anchored to the ground. Without any means of flexibility or escape, the building is a sitting target. A building's massive size offers a greater surface area to the wind, allowing greater total force for the same wind pressure than a tree could offer.

But a tree in a hurricane is like Muhammad Ali doing the rope-a-dope. The tree is going to be hit by the hurricane, but it yields and turns and shuffles its way until the force of the hurricane no longer threatens it. In this case, the metaphor is literal: by bending with the wind, the tree and its leaves can sometimes escape totally unscathed.

Richard A. Anthes, director of the National Center for Atmospheric Research, offers another reason why we see so many buildings, and especially so many roofs, blown away during a hurricane. "Buildings offer a surface which provides a large aerodynamic lift, much as an airplane wing. This lift is often what causes the roof to literally be lifted off the building."

DAVID FELDMAN

We don't want to leave the impression that trees can laugh off a hurricane. Many get uprooted and are stripped of their leaves. Often we get the wrong impression because photojournalists love to capture ironic shots of buildings torn asunder while Mother Nature, in the form of a solitary, untouched, majestic tree, stands triumphant alongside the carnage.

Submitted by Daniel Marcus of Watertown, Massachusetts.

Why Are Downhill Ski Poles Bent?

Unlike the slalom skier's poles, which must make cuts in the snow to negotiate the gates, the main purpose of the downhill ski poles is to get the skier moving, into a tuck position . . . and then not get in the way.

According to Tim Ross, director of Coaches' Education for the United States Ski Coaches Association, the bends allow the racer "to get in the most aerodynamic position possible. This is extremely important at the higher speeds of downhill." Savings of hundredths of a second are serious business for competitive downhill skiers, even when they are attaining speeds of 60–75 miles per hour.

If the bends in the pole are not symmetrical, they are designed with careful consideration. Dave Hamilton, of the Professional Ski Instructors of America, reports that top-level ski racers have poles individually designed to fit their dimensions. Recreational skiers are now starting to bend their poles out of shape. According to Ross, the custom-made downhill ski poles may have as many as three to four different bend angles.

Funny. We haven't seen downhill skiers with three to four different bend angles in their bodies.

Submitted by Roy Welland of New York, New York.

Why Do So Many Mail-Order Ads Say to "Allow Six to Eight Weeks for Delivery"? Does It Really Take that Long for Companies to Process Orders?

This is a mystery we have pondered over ourselves, especially since these same companies that warn us of six-to-eight-week delivery schedules usually send us our goods within a few weeks. We talked to several experts in the mail-order field who assured us that any reasonably efficient operation should be able to ship items to customers within two to three weeks.

Many manufacturers farm out much or all of the processing of mail orders to specialized companies, called fulfillment houses. Some fulfillment houses do everything from receiving the initial letters from customers and obtaining the proper goods from their own warehouses to producing address labels, maintaining inventory control, and shipping out the package back to the customer.

DAVID FELDMAN

Dick Levinson, of the fulfillment company H.Y. Aids Group, told *Imponderables* that a fulfillment house should be able to gurarantee a client a turnaround of no more than five days from when a check is received until the package is shipped to the customer. A two- or three-day turnaround is the norm.

Do the mail order companies blame the post office? Why not? Everybody else does. But despite a few carpings, all agreed that even third-class packages tend to get delivered anywhere in the continental United States within a week.

Being paranoid types, we thought about a few nefarious reasons why mail-order companies might want to delay orders. Perhaps they want to create a little extra cash flow by holding on to checks for an extra month or so? No, insisted all of our sources.

How about advertising goods they don't have in stock? As checks clear, companies could pay for their inventory out of customer money rather than their own. It's possible but unlikely, said our panel. Stanley J. Fenvessey, founder of Fenvessey Consulting and perhaps the foremost expert on fulfillment, said that only a fly-by-night operation would try to get away with such shenanigans. He offered a few more benign explanations.

Sometimes a mail-order company, particularly one that specializes in imports or seasonal items, might run out of stock temporarily. By listing a delayed delivery date, the company forestalls complaints, even though it expects to deliver merchandise in half the stated time.

And in the magazine field, fledgling efforts sometimes try a "dry test," in which prospective subscribers are solicited by mail even though no magazine yet exists. Only if there is a high enough response rate will the magazine ever be produced.

The most compelling reason is the Federal Trade Commission's Mail Order Rule. The rule was established in 1974 after consumers complained in droves about late or nonexistent shipments of merchandise by mail-order operations. The President's Office of Consumer Affairs reported that the number of complaints registered against mail-order firms was second only to complaints about automobiles and auto services.

The Mail Order Rule states that a buyer has the right to

assume that goods will be shipped within the time specified in a solicitation and, "if no time period is clearly and conspicuously stated, within thirty days after receipt of a properly completed order from the buyer." Furthermore, when a seller is unable to ship merchandise within the time provisions of the rule, the seller must not only notify the buyer of the delay but also offer the option to the buyer to cancel the order.

Refunding money is not exactly any company's favorite thing to do, but the provisions about sending the notice of delay and option to cancel is perhaps more onerous to mail-order firms. Not only must the seller spend money on mailing these notices, but must somehow track the progress of each order to make sure it hasn't exceeded the 30-day limit. The bookkeeping burden is enormous.

Finally, we have arrived at the answer: By putting a shipping deadline of much longer than they think they will ever need, mail-order firms avoid having to comply with the provisions of the thirty-day rule whenever they run out of stock temporarily.

But don't these disclaimers discourage sales? After all, most items ordered by mail are available in retail stores as well. Dick Levinson suggests that most items ordered from magazines and newspapers are impulse items rather than necessities, and that most buyers are flexible about delivery schedules. Lynn Hamlin, book buyer for New York's NSI Syndications Inc., commented that space customers (those who order from newspapers and magazines) are less demanding than those who order from catalogs with toll-free phone numbers and who have the ability to ask a company operator how long the delivery will take. NSI advertisements guarantee shipment within 60 days, but usually are filled in two or three weeks. Ms. Hamlin notes that she has not seen any detrimental effect of the sixty-day guarantee on her company's sales, although she admits that around December 1, some potential customers might fear whether merchandise would arrive by Christmas.

Stanley Fenvessey informs us that about 75 to 90% of all catalog merchandise is delivered within two weeks, and insists

DAVID FELDMAN

that no large catalog house would ever print "six to eight weeks for delivery." One of Fenvessey's smaller clients, who owned a catalog company, printed "please allow four to five weeks for delivery" on his catalog. Fenvessey asked his client whether it really took this long to fulfill orders. The client replied that most orders were delivered in two weeks.

"So why put four to five weeks in the catalog?" asked Fenvessey.

"Because this way we avoid hassles when we are a few days late."

Fenvessey was convinced that the client couldn't see the forest for the trees. Fenvessey conducted a test in which two sets of catalogs were printed and shipped; the only difference between the two was that one announced that delivery would be between two to three weeks; the other, four to five weeks. The two to three week catalog drew 25% more orders, a huge difference.

Maybe many space advertisers are losing sales by scaring potential customers into thinking they're going to have to wait longer than they really will to get merchandise.

Submitted by Susie T. Kowalski of Middlefield, Ohio.

Why Are Silos Round?

The poser of this Imponderable, Susan Diffenderffer, insisted she had the correct answer in hand: "In a square silo, grain could form an air pocket and cause spontaneous combustion. There are no corners in a round silo."

Well, we think the spontaneous combustion theory is a tad apocalyptical, but you have the rest of the story right. Actually, at one time silos were square or rectangular. Fred Hatch, a farmer from Illinois, built a square wooden silo in 1873. But the square corners didn't allow Hatch to pack the silo tightly enough. As a result, air got in the silo and spoiled much of the feed. To the rescue came Wisconsin agricultural scientist, Franklin H. King, who built a round silo ten years later. The rest is silage history.

Why is it so important to shut air out of a silo? The mold that spoils grain cannot survive without air. Without air, the grass and corn actually ferment while in the silo, inducing a chemical change in the silage that makes it palatable all through the winter season.

Before silos were invented, cows gave less milk during winter because they had no green grass to eat. Silos gave the cows the lavish opportunity to eat sorghums all year long.

Submitted by Susan C. Diffenderffer of Cockeysville, Maryland.

DAVID FELDMAN

Why Does Dialing 9 Usually Get You an Outside Line in a Hotel? And Why Does 8 Open a Long-Distance Line?

For many years we've been looking at want ads in the newspaper and seeing positions open for PBX operators. We've always wondered what the heck they did. "PBX" sure sounds threateningly high-tech. Little did we know that we were already experts in the field.

PBX systems are simply telephone lines designed for communication within one building or business that are also capable of interfacing with the outside world. Most large hotels have a PBX system. When you lift your phone up in your room, you become a PBX station user whether you like it or not.

Most PBX systems reserve numbers one through seven for dial access to other internal PBX stations. In a hotel, this allows a guest in one room to call another room directly. Decades ago, one might have dialed for the operator to perform this function, but hotels found that patrons preferred the greater speed of direct access; and of course, direct dialing saved hotels the labor costs of operators.

There is no inherent reason why 4 or 2 *couldn't* be the access code for an outside line or long-distance access, but Victor J. Toth, representing the Multi-Tenant Telecommunications Association, explains how the current practice began:

> The level "9" code is usually used by convention in all commercial PBX and Centrex as the dialing code for reaching an outside line. This number was chosen because it was usually high enough in the number sequence so as not to interfere with a set of assigned station numbers (or, in the case of a hotel, a room number).

Likewise, the 8 is sufficiently high in the number sequence to not interfere with other station numbers and has become the conventional way to gain access to long-distance services.

Toth adds that it is easy to deny level 9 class of service to a particular phone or set of phones if desired. Most hotels, for example, make it impossible for someone using a lobby phone to dial outside the hotel, let alone long distance.

Why Can't (Or Won't) Western Union Transmit an Exclamation Mark in a Telegram?

Many of the origins of the customs we now take for granted are lost in obscurity. We are thankful to Paul N. Dane, executive director of the Society of Wireless Pioneers, who led us to two gentlemen, W.K. "Bill" Dunbar, and Colonel Ronald G. Martin, who could answer these two Imponderables authoritatively.

Mr. Dunbar informs us that the original Morse code alphabet (but not the international code used for cablegrams and radiograms) did indeed provide for the exclamation mark: - - - · expressed it. "The early teletype machines with a three-row keyboard may not have provided for the exclamation mark, and although later equipment did, it might not have been capable of conveying the exclamation point into a Telex circuit."

DAVID FELDMAN

According to Colonel Martin:

> It is very easy to cause an error during the transmission of a message with a lot of punctuation therein. Therefore, Western Union, in order to prevent lawsuits, abolished it.

Even if there were technological problems in printing an exclamation mark, a more compelling reason existed to shun it and other punctuation: Punctuation marks were charged as if they were words.

Submitted by Fred T. Beeman of Wailuku, Hawaii.

Why Do Telegrams End Sentences with STOP Rather than with a Period?

Western Union, throughout most of its history, has charged extra for periods as well as exclamation marks. But the reasons for the exclusion of periods and the inclusion of STOP are fascinating and highly technical. Bill Dunbar, president of the Morse Telegraph Club, explains:

> In certain instances the word STOP, when used as a period, was free. I believe this was the case with transoceanic cablegrams. Hollywood sometimes showed STOP in domestic telegrams, which may have given the impression it was common usage. At one time when competition between Western Union and Postal Telegraph was keen, STOP was free, but this did not last long and usually it was a chargeable word, so naturally it wasn't used much.
>
> The main reason for periods not appearing was a procedural one—the period was used to indicate the beginning and end of the body of a message. The preamble (i.e., call letters of sending office, the number of words in a message, type of service, and type of payment), origin city, the time and date were sent first. This was followed by the word TO, after which the receiving operator would drop down a line or two and move to the left of the page to write the address.

At the end of the address, a period was sent, signifying that the next characters would begin the text of the message. At the end of the message, the sender would send another period and say SIG (signature), and the copying operator would drop down two lines to write the signature; he would also add the time the message was copied. If there *was* a period in the message, it was converted to STOP for transmission.

The words TO, SIG and the periods were not written on the telegram, since they were procedural signals. Decimals were transmitted by sending the word DOT as 18 DOT 5. This might seem clumsy, but it eliminated any ambiguity as to whether a decimal point or the end of a message was indicated.

PERIOD must have been considered instead of STOP to signify the period, but probably was rejected for one simple reason: STOP is two letters shorter. Colonel Martin adds that the word PERIOD is more likely to cause confusion when a telegram concerns time.

Both Martin and Dunbar emphasize how important brevity of language and speed of transmission has always been to Western Union. But customers have proven to be just as frugal in their own way. Traditionally, Western Union charged a basic rate that allowed for ten free words. Any extra words or punctuation marks cost extra. Sometimes the need to squeeze a lot of information into ten words tested the ingenuity of the sender, as Mr. Dunbar's story illustrates:

> The story is told of a man who sent the following message: BRUISES HURT ERASED AFFORD ERECTED ANALYSIS HURT TOO INFECTIONS DEAD
> Translated it reads: "Bruce is hurt he raced a Ford he wrecked it Aunt Alice is hurt too in fact she's dead"

Writers who are paid by the word try to be as verbose as possible. But a writer who has to *pay* by the word will try to squeeze nineteen words into ten.

Submitted by Eileen LaForce of Weedsport, New York.

DAVID FELDMAN

Why Are Most Snack-Food Items, Such as Chips, Cakes, and Popcorn, Prepriced (on the Package) by the Manufacturers?

How often have you scoured the aisles of your local supermarket looking for the elusive item on your grocery list? You despair of ever finding what you need when you encounter a young man arduously arranging packages on the shelf. "Where can I find the artificial coloring?" you inquire.

"I don't know. I don't work here," replies the man.

Why can't you ever find the people who supposedly *do* work at the damn store? This poignant episode, repeated in grocery stores throughout the land, explains—believe it or not—why most snack items are prepriced by the manufacturers.

Most items in a supermarket, such as canned goods, are sent to the store by a warehouse distributor who handles many different brands. Snack-food manufacturers work on "store-door distribution," providing full service to retailers. Potato chips or popcorn are brought to the stores in trucks displaying the logo of one company. The agent for the manufacturer rids the shelves of any unsold packages with elapsed expiration dates, restocks, and straightens up the shelves to make the company's selling environment look attractive.

Retailers have come to expect this kind of full-service treatment from the snack-food industry. Next to the expense of cashiers, pricing items is one of the costliest labor costs of grocery retailers: Stores welcome prepricing by the industry.

Why do snack-food manufacturers go along with providing extra service to stores? Although manufacturers like retaining the control of pricing, according to Chris Abernathy, of the Borden Snack Group, fear of retail overcharging is not the main purpose for the practice. By stamping the price themselves, Borden and other snack-food companies can run citywide or regional promotions by cutting the price on the package itself.

Al Rickard, of the Snack Food Association, stresses that by stamping prices on packages themselves, manufacturers can guarantee *equality* of prices to outlets that sell their products. Snack foods are sold not only in grocery stores but in convenience stores, bowling centers, service stations, and other venues that are not used to putting price stickers on food items. Those establishments are more likely to sell snacks when they don't feel they will be undercut in price by supermarkets.

Most important, with store-door distribution manufacturers can assure themselves that their products are not languishing on the shelves because retailers are refusing to pull old goods. What all of the food items with prepricing have in common is their perishability. Most salted snack foods have shelf lives of approximately two weeks. Other prepriced items, such as doughnuts and bread, may have even shorter expiration dates. If they have to preprice snack items to guarantee the proper rotation of their goods, it is a small price to pay.

Submitted by Herbert Kraut of Forest Hills, New York.

DAVID FELDMAN

Why Are the Commercials Louder than the Programming on Television?

Having lived in apartments most of our adult lives, we developed a theory about this Imponderable. Let us use a hypothetical example to explain our argument.

Let's say a sensitive, considerate yet charismatic young man —we'll call him "Dave"—is taking a brief break from his tireless work to watch TV late at night. As an utterly sympathetic and empathic individual, "Dave" puts the volume at a low level so as not to wake the neighbors who are divided from him by tissue-thin walls. Disappointed that "Masterpiece Theatre" is not run at 2:00 A.M., "Dave" settles for a rerun of "Hogan's Heroes." While he is studying the content of the show to determine what the character of Colonel Klink says about our contemporary society, a used-car commercial featuring a screaming huckster comes on at a much louder volume.

What does "Dave" do? He goes up to the television and

lowers the volume. But then the show comes back on, and "Dave" can't hear it. Ordinarily, "Dave" would love to forgo watching such drivel, so that he could go back to his work as, say, a writer. But he is now determined to ascertain the sociological significance of "Hogan's Heroes." So for the sake of sociology, "Dave" gets back up and turns the volume back on loud enough so that he can hear but softly enough not to rouse the neighbors. When the next set of commercials comes on, the process is repeated.

Isn't it clear? Commercials are louder to force couch potatoes (or sociological researchers) to get some exercise! When one is slouched on the couch, the walk to and from the television set constitutes aerobic exercise.

Of course, not everyone subscribes to our theory.

Advertising research reveals, unfortunately, that while commercials with quick cuts and frolicking couples win Clio awards, irritating commercials sell merchandise. And it is far more important for a commercial to be noticed than to be liked or admired. Advertisers would like their commercials to be as loud as possible.

The Federal Communications Commission has tried to solve the problem of blaring commercials by setting maximum volume levels called "peak audio voltage." But the advertising community is way ahead of the FCC. Through a technique called "volume compression," the audio transmission is modified *so that all sounds, spoken or musical, are at or near the maximum allowable volume.* Even loud rock music has peaks and valleys of loudness, but with volume compression, the average volume of the commercial will register as loudly as the peaks of regular programming, without violating FCC regulations.

The networks are not the villain in this story. In fact, CBS developed a device to measure and counterattack volume compression, so the game among the advertisers, networks, and the FCC continues. Not every commercial uses volume compression, but enough do to foil local stations everywhere.

Of course, it could be argued that advertisers have only the

DAVID FELDMAN

best interests of the public at heart. After all, they are offering free aerobic exercise to folks like "Dave." And for confirmed couch potatoes, they are pointing out the advantages of remote-control televisions.

Submitted by Tammy Madill of Millington, Tennessee.
Thanks also to Joanne Walker of Ashland, Massachusetts.

Why Is U.S. Paper Money Green When Most Countries Color-Code Their Currency?

Until well into the nineteenth century, paper money was relatively rare in the United States. But banknotes became popular in the mid-1800s. These bills were printed in black but included colored tints to help foil counterfeiters.

However, cameras then in existence saw everything in black, rendering color variations in bills meaningless when reproduced photographically. According to the U.S. Treasury, the counterfeiters took advantage:

> the counterfeiter soon discovered that the colored inks then in use could easily be removed from a note without disturbing the black ink. He could eradicate the colored portion, photograph the remainder, and then make a desired number of copies to be overprinted with an imitation of the colored parts.

Tracy R. Edson, one of the founders of the American Bank Note Company, developed the solution. He developed an ink that could not be erased without hurting the black coloring. Edson was rewarded for his discovery by receiving a contract from the U.S. government to produce notes for them. Edson's counterfeit-proof ink had a green tint.

In the nineteenth century, notes were produced by private firms as well as the treasury. But all notes, regardless of where they were printed, were issued in green, presumably to provide uniformity.

Could Edson have chosen blue or red instead of a green tint? Certainly. Although our sources couldn't tell us why green was the original choice, the treasury does have information about why the green tint was retained in 1929, when small-sized notes were introduced:

> the use of green was continued because pigment of that color was readily available in large quantity, the color was relatively high in its resistance to chemical and physical changes, and green was psychologically identified with the strong and stable credit of the Government.

And besides, "redbacks" or "bluebacks" just don't have a ring to them.

Other countries vary the coloring of their bills as well as their size. And why not? Different sizes would enable the sighted but especially the legally blind to sort the denominations of bills easily. But despite occasional rumblings from legislators, the Treasury Department stands by its greenbacks.

Submitted by Paul Stossel of New York, New York. Thanks also to Charles Devine of Plum, Pennsylvania; and Kent Hall of Louisville, Kentucky.

Why Do We Have to Close Our Eyes When We Sneeze?

We thought we'd get off easy with this mystery. Sure, a true Imponderable can't be answered by a standard reference work, but would a poke in a few medical texts do our readers any harm?

We shouldn't have bothered. We understand now that a sneeze is usually a physiological response to an irritant of some sort. We learned that there is a $10 word for sneezing (the "sternutatory reflex") and that almost all animals sneeze. But what exactly happens when we sneeze? Here's a short excerpt from one textbook's explanation of a sneeze:

DAVID FELDMAN

When an irritant contacts the nasal mucosa, the trigeminal nerve provides the affect limb for impulses to the pons, and medullai Preganglionic efferent fibers leave these latter two structures via the intermediate nerve, through geniculate ganglion to the greater petrosal nerve, through the vividian nerve and then synapse at the sphenopalatine ganglion . . .

Get this outta here! Until Cliff Notes comes out with a companion to rhinology textbooks, we'll go to humans for the answers.

Our rhinologist friend, Dr. Pat Barelli, managed to read those textbooks and still writes like a human being. He explains that the sneeze reflex is a protective phenomenon:

The sneeze clears the nose and head and injects O_2 into the cells of the body, provoking much the same physiological effect as sniffing snuff or cocaine. When a person sneezes, all body functions cease. Tremendous stress is put on the body by the sneeze, especially the eyes.

As Dr. G. H. Drumheller, of the International Rhinological Society, put it, "we close our eyes when sneezing to keep the eyes from extruding." While nobody is willing to test the hypothesis, there is more than a grain of truth to the folk wisdom that closing your eyes when you sneeze keeps them from popping out, but probably not more than three or four grains.

Submitted by Linda Rudd of Houston, Texas. Thanks also to Michelle Zielinski of Arnold, Missouri; Helen Moore of New York, New York; Jose Elizondo of Pontiac, Michigan; Amy Harding of Dixon, Kentucky; and Gail Lee of Los Angeles, California.

Why Don't Grazing Animals that Roll in or Eat Poison Ivy Ever Seem to Get Blisters or Itching in Their Mouths?

A few of the many veterinarians we spoke to had seen allergic reactions to poison ivy among animals but all agreed it was exceedingly rare. Poison ivy is not really poison. Humans develop an allergic reaction because of a local hypersensitivity to the oil in the plant. Veterinarian Anthony L. Kiorpes, a professor at the University of Wisconsin-Madison School of Veterinary Medicine, informed us that the same plant that may cause a severe reaction in one human may not affect another person at all.

Elizabeth Williams, of the University of Wyoming College of Agriculture, notes that she has never seen an allergic reaction in deer, but allows:

> It's possible some deer might be allergic to it but we just don't see the reaction because they are covered with hair. Or it may be that only a very few deer are allergic, and they learn to stay away from poison ivy.

DAVID FELDMAN

Veterinarian Ben Klein feels that most domestic animals have a built-in immunity to contact allergy dermatitis, such as poison ivy. Furthermore, that same hair Dr. Williams mentioned hiding an allergic reaction also shields the skin against potential reactions, according to veterinary dermatologist Peter Ihrke.

Why don't ruminants break out when they eat poison ivy or poison oak? Dr. Don E. Bailey, secretary-treasurer of the American Association of Sheep and Goat Practitioners, explains that even if these animals had a tendency toward allergic reactions, which they don't, the mucus membrane in their mouths is very thick and heavy.

The one animal that most often seems to contract allergic reactions to poison ivy is the dog. Dogs love to roll around in the worst imaginable things. Dr. Ihrke notes that most dogs can withstand the exposure to poison ivy but many of their owners cannot. The owners pet the dogs and come down with severe reactions. Similarly, an innocent vet will examine a dog and break out in a rash, the victim of a communicable disease that doesn't afflict the carrier.

Submitted by Karole Rathouz of Mehlville, Missouri.

Why Don't Queen-Sized Sheets Fit My Queen-Sized Bed?

Queen-sized beds expanded from 60" x 75" to 60" x 80" in the early 1960s. You would think that more than twenty-five years would be a sufficient amount of time to manufacture sheets large enough to cover the expanded surface area. And it was.

The problem is that sheets are designed to cover mattresses, and the linen industry has no control over what the mattress manufacturers are doing. And what the bed companies are doing lately is driving sheetmakers nuts. As Richard Welsh, senior vice president of Cannon Mills Company, succinctly summarizes:

The sheet industry has experienced problems with fitted sheets for all sizes, not only queen size. The problem is primarily due to the fact that mattress manufacturers have been increasing the depth of their mattresses. As one tries to get an edge on the other, they outsize them by half an inch. There are no standard mattress depths.

When Mr. Welsh first wrote to *Imponderables,* in December 1986, he complained about how the sheet industry, accustomed to fitting six-and-one-half- to seven-inch-deep mattresses, watched in horror as depth inflation hit. In the early 1980s, Cannon increased the length of their sheets to accommodate mattresses from eight to eight-and-one-half inches deep. But soon, the nine-inch barrier was broken. Cannon responded to this problem in 1987 by manufacturing sheets "guaranteed to fit." These sheets could cover a mattress nine-and-one-half inches thick.

But the mattressmakers never stopped. They invented a whole new genre of bed, the "pillow top" mattress, with pockets of polyester fill on top. Pillow tops have increased the crown space (the highest point of padding) on some mattresses to as high as one foot to twelve-and-one-half inches.

The standard queen-size flat sheet is now 90" x 102", which is more than sufficient to cover the 60" x 80" queen-size beds if they don't continue the creep upward. But creep they probably will. On the low end of the market, six-inch-thick mattresses still are available. If you have sheets that are longer than you need, have compassion for the sucker with the pillow top. The same sheet that is too long for you probably isn't long enough to tuck under his mattress.

P.S. It was inevitable. Fieldcrest and Wamsutta, among others, are now manufacturing sheets specifically for pillow top mattresses.

Why Is There Cotton Stuffed in Prescription and Over-the-Counter Medicine Bottles? What Happens If I Take Out the Cotton? Why Aren't Alka-Seltzer Containers Stuffed with Cotton?

The main purpose of the cotton stuffed in medicine vials is to prevent rattling and subsequent breakage of pills during shipment. But why cotton? Because of its absorbency, cotton helps keep medications dry. David G. Miller, associate director of the National Association of Retail Druggists, points out that moisture will destroy most drugs.

Still, all the druggists we spoke to recommended taking out the cotton once the container is opened for use. Melvin T. Wilczynski, of the Lane Drug Company, explains that the absorptive characteristics of cotton, which help keep pills dry during shipment, also are capable of absorbing moisture from the environment. If the cotton gets wet and re-enters the bottle, the effectiveness of the medication is jeopardized.

WHEN DO FISH SLEEP?

Excess heat and light are also capable of breaking down medications. For this reason, it makes no sense to keep pills in the kitchen, where they are exposed to the heat of ovens, or outside the medicine cabinet in the bathroom, where they could be exposed to harsh light or space heaters.

At one time, Miles Laboratories did put cotton into Alka-Seltzer containers, but found that consumers couldn't be trusted. If you accidentally get an Alka-Seltzer tablet wet, you get premature fizz—a temporary thrill perhaps, but one that will do you no good when a bout of indigestion sets in.

Miles provides a styrofoam cushion to protect tablets during shipping, but they recommend throwing away the cushion once the bottle is opened. Although styrofoam is not as absorbent as cotton, it is perfectly capable of generating bubbles when wet.

Submitted by Andrew Neiman of Dallas, Texas.

Why Do Bagels Have Holes?

In *Why Do Clocks Run Clockwise? and Other Imponderables*, we explained why doughnuts have holes. We were pretty smug about our accomplishment too. Then a letter arrives from Jay Howard Horne asking us why bagels have holes. Will there ever be a stop to this mania for knowledge about hole origins?

Nobody knows for sure who created the first bagel. Chances are, it was an accident precipitated by a piece of yeast-laden dough falling into hot water. But we do know who first called a bagel a "bagel." In 1683, the first Viennese coffeehouse was opened by a Polish adventurer, who introduced a new bread called the *beugel*. When Austrians emigrated to the United States in the next two centuries, the beugel was re-christened the bagel.

So what was a Polish man doing opening a coffeehouse in Vienna and creating a hole-y bread?

DAVID FELDMAN

The king of Poland, Jan Sobiesky, had become a hero in Austria in the late seventeenth century by driving off armed invaders from Turkey. In their escape, the Turks left behind sacks of enough coffee to keep every citizen of Vienna up nights for a month, inspiring the opening of many a coffeehouse in Vienna.

The coffeehouse owner took a popular yeast bread called kipfel and reshaped it into the bagel shape we know and love today. The bread was meant to resemble the stirrups of brave King Sobiesky, who fought on horseback to save Vienna from the Turks. "Bagel" is derived from the German word for stirrup, "bugel."

Submitted by Jay Howard Horne of Pittsburgh, Pennsylvania.

Do the Digits in a Social Security Number Have Any Particular Meaning?

Now that the Social Security number has become a virtual citizenship identification number, paranoid types have become convinced that each digit is another way for Uncle Sam to poke into our private lives. No, the government can't tell by looking at our Social Security number whether we are registered Democrats or Republicans, whether we are in the highest income-tax bracket or are on welfare, or even whether we have committed a crime.

Under the current system, the first three digits of a Social Security number indicate the state of residence of the holder at the time the number was issued. The remaining digits have no special meaning.

Before 1973, Social Security numbers were assigned by local Social Security offices. The first three digits were assigned based on the location of the Social Security office rather than the residence of the issuee. Opportunists used to scoop up several different Social Security numbers by applying for cards at several different offices, which led to the current practice of issuing all numbers from the central Social Security office in Baltimore. According to Dorcas R. Hardy, commissioner of Social Security, the first three digits of a person's Social Security number are now determined by the ZIP code of the mailing address shown on the application for a Social Security number.

Although the first three digits of the Social Security number do not correspond exactly to the first three digits of that state's zip codes, the lowest Social Security numbers, like their ZIP code counterparts, start in New England and then get progressively larger as they spread westward. Numbers 001–003 are assigned to New Hampshire, and the highest numbers assigned to the 50 states are New Mexico's 585. The Virgin Islands (580), Puerto Rico (580–584, 596–599), Guam (586), American Samoa (586), and the Philippine Islands (586) are also assigned specific three-digit codes.

Until 1963, railroad employees were issued a special series of numbers starting with the digits 700–728. Although this practice is now discontinued, these numbers remain the highest ever issued. No one has ever cracked the 729 plus barrier.

*Submitted by Douglas Watkins, Jr. of Hayward, California.
Thanks also to Jose Elizondo of Pontiac, Michigan; Kenneth
Shaw of San Francisco, California; and Rebecca Lash of Ithaca,
New York.*

DAVID FELDMAN

Why Do the Light Bulbs in My Lamps Loosen After I've Put Them in Place?

An unscientific poll conducted by the Imponderables Research Board indicates that creeping bulb loosening is a problem for many, although a majority of respondents never faced the problem. Is some sadist running around loosening the bulbs of selected victims?

Perhaps, but a natural explanation is more likely. The greatest culprit in loosening light bulbs is vibration. Friction keeps the socket threads of a light bulb tightly fitted into the base threads of a fixture. J. Robert Moody, of General Electric, informed Imponderables that "vibration weakens the friction force, allowing the light bulb to back out of the socket on its own. If the vibration is intense, like on an automobile or an airplane, then a bayonet base must be used in place of the screw-threaded base."

Perhaps that incessant bass drone emanating from the

heavy-metal freak upstairs caused your problem. The only solution might be the purchase of a bayonet base for your lamp or a bayonet to use on your neighbor.

Submitted by Darryl Williams of New York, New York.

HOW Are Olives Pitted? How Do They Stuff Olives?

Until recently, the vast majority of olives were stuffed by hand. Olives were held in cups, and a crude machine operated with a foot treadle would punch out the pit while another element cut a hole on top of the olive simultaneously. A worker would then inspect the olive. If it was acceptable, she would take a pimento, onion, anchovy or other filling and manually stick it in the hole.

Obviously, olive companies were desperately in need of a high-tech solution to the slowness of their production line. Not only was the pitting and stuffing operation labor-intensive, but the machines would rip olives to shreds and leave pit fragments as a "bonus" for unsuspecting consumers. Even more damaging, the U.S. Food and Drug Administration would routinely refuse to allow importation of mangled olives (almost all green olives are imported from Spain).

Automation revolutionized the olive industry in the early 1970s. Modern machines, typically containing twenty-four separate stations, are capable of stuffing twelve hundred to fifteen hundred olives a minute. The olives are pitted in one movement, and the pimento is inserted with ease.

The down side to this otherwise lovely story is that automation has encouraged olive distributors to dump natural pimentos in favor of pimentos "enhanced" with paste and binders. These additives enable the pimento to be fashioned into an endless

DAVID FELDMAN

ribbon of red stuff. The machine then cuts the ribbon to exact specifications prior to stuffing the cavity, so that larger olives receive wider strips of pimento—red stuff. Of course, the pimento—red stuff tastes more like red stuff than pimento, but this is the price we pay for progress.

Machines now exist to sort olives by size, to inject brine into a jar, to pack olives in jars, to stuff olives with pimentos, to slice olives, and to seal olive jars. Until the 1980s almost all olives were packed into jars by hand. Fancy Spanish olives are often placed in geometric patterns to induce impulse purchases by consumers. According to Edward Culleton, of the Green Olive Trade Association, American consumers have never developed brand loyalty, so shoppers have traditionally been receptive to eye-catching arrangements of olives. About 90% of green olives are now packed by machine rather than by hand, so "place packs" (hand-packed jars) are likely to be a specialty item in the future. In fact, Spain now exports hand-pitted stuffed olives in beautiful crystal jars as a luxury gift item—and the olives are stuffed with real pimentos.

Submitted by Helen Tvorik of Mayfield Heights, Ohio.

Why Is One Side of a Halibut Dark and the Other Side Light?

With the price of halibut these days, we might assume that we are paying extra for the two-tone job. But nature supplies halibut with two colors for a less mercenary reason.

The eyeless side of the halibut is light, requiring no camouflage. But the side with eyes is dark. Like other flat fish that swim on one side, the halibut is dark on the side exposed to the light. Robert L. Collette, associate director of the National Fisheries Institute, describes the coloring as "a natural defense system." The dark side is at top "so that predators looking down upon halibut are less likely to detect their presence."

This camouflage system is adapted for fish and mammals that swim upright. They have dark backs and white undersides to elude their predators.

What Is the Difference Between a "Mountain" and a "Hill"?

Although we think you are making a mountain out of a molehill, we'll answer this Imponderable anyway. Most American geographers refer to a hill as a natural elevation that is smaller than 1,000 feet. Anything above 1,000 feet is usually called a mountain. In Great Britain, the traditional boundary line between hill and mountain is 2,000 feet.

Still, some geographers are not satisfied with this definition. "Hill" conjures up rolling terrain; "mountains" connote abrupt, peaked structures. A mound that rises two feet above the surrounding earth may attain an elevation of 8,000 feet, if it happens to be located in the middle of the Rockies, whereas a 999-foot elevation, starting from a sea-level base, will appear massive. For this reason, most geographers feel that "mountain" may be used for elevations under 1,000 feet if they rise abruptly from the surrounding terrain.

WHEN DO FISH SLEEP?

The *Oxford English Dictionary* states that "hill" may also refer to non-natural formations, such as sand heaps, mounds, or, indeed, molehills.

Submitted by Thomas J. Schoeck of Slingerlands, New York.
Thanks also to F. S. Sewell of San Jose, California.

Why Aren't There License Plates on the Back of Many Big Trucks on the Highway?

Fewer than two-thirds of the fifty states require license plates on both the front and back of a commercial truck. Why do truckers, unlike automobile owners, only have to display one plate in many states?

Presumably, tractors will be pulling trailers most of the time, so the only time we are likely to see a tractor with two plates is when it is "deadheading" (not towing a trailer). Then, the back license plate is likely to be obscured by the trailer anyway, and be of little use to police.

Because many tractors are crossing borders constantly, the licensing of commercial tractors and trailers can be complicated. According to Jan Balkin, of the American Trucking Associations,

> All trailers must have license plates from the state in which it is licensed. That state may not necessarily be the same as the state in which the tractor is licensed; carriers may license the tractor and trailer in different states, depending upon certain financial decisions as to which state(s) the carrier chooses.

DAVID FELDMAN

Why Do Mayors Hand Out Keys to Their City?

We've all seen those silly ceremonies on TV where a grinning mayor hands a three-foot-long key to a minor celebrity as flash-bulbs pop. But we have always wondered: Why does the recipient need a key to the city? He's already *in* the city.

Actually, this ceremony has legitimate historical antecedents. In the Middle Ages, most large cities were walled. Visitors could enter and exit only through gates that were locked at sundown and reopened at dawn.

Mike Brown, of the United States Conference of Mayors, told *Imponderables* that gatekeepers used keys to open and close the gates. These keys were closely guarded, for they were crucial in preventing military attacks. If a key was passed to an honored visitor, it indicated total trust in him.

Today, a mayor no longer threatens the security of her domain by handing out the key to the city, and the honor is more likely a public relations stunt than in gratitude for service or accomplishment. But the meaning is the same. By handing out the key to the city, the mayor says, "Come back any time and you don't even have to knock. We trust you."

What Is the Purpose of the Beard on a Turkey?

All of our poultry experts felt that the beard has no specific anatomical function, but this doesn't mean the beard has no purpose. The beard is a secondary sex characteristic of the male, a visual differentiation between the sexes. How could a hen possibly resist the sexual allure of the beard of a strutting Tom?

Submitted by Mrs. Anabell Cregger of Wytheville, Virginia.

WHY Are Banking Hours So Short?

Nine to three, five days a week. Not a bad job if you can get it, eh? Short banking hours have always fit the needs of bankers and industry, but have not been convenient for retail customers. The banks are open only when the average person is working or going to school.

Mind you, workers in the bank industry don't get to leave the door at the stroke of three. Tellers, for example, must count cash and report their balances to a central processing center. Before automation, reconciling their books might have taken a little longer, but not significantly so.

Executives in the banking industry, who do not have to mind the day-to-day transactions of retail customers, have plenty of phone and social contacts to make after banking hours. When we posed this Imponderable to Joan Silverman, of Citibank, she was incredulous that we had not heard of the rule of 3/6/3.

"What is the rule of 3/6/3?" we asked.

"It's simple," she replied. "If you want to be a successful

DAVID FELDMAN

banker, you pay 3% to depositors; you charge 6% on loans to customers; and you hit the golf course by 3 P.M." Based on the interest rates mentioned, this obviously is a very old rule.

Government regulations once restricted the hours during which banks could be open for business. Gentleman bankers conducted most of their business on the golf course while tellers were back at the ranch settling their ledgers.

With computerization, there is no reason why banks couldn't have much longer hours. The tradition of the 9 A.M. to 3 P.M. Monday through Friday bank is preserved because of the bottom line: The banks figure that if they stay open later or open on weekends, they will increase retail customers' simple—and to the bank, often unprofitable—transactions, such as depositing or withdrawing money from checking and savings accounts. Most other businesses are closed on weekends and evenings—*they* don't have demanding, long hours.

Bank hours generally are extended only when there is a competitive marketing reason to do so, usually when a new bank or new branch needs to build new accounts and can advertise extended hours. Ohio's Banc One, Dayton, for example, opens many branches on Saturdays and even Sundays. Many of their branches are located in malls; before Banc One opened on weekends, it was often the only business closed in the whole shopping center. Much to Banc One's surprise, according to *American Banker:*

> the volume of teller activity during Sunday's four-hour shift has been greater than the amount of teller activity during a normal seven-hour weekday.

Obviously, if it were cost-effective for most banks to be open longer, they would do it. Automated teller machines have effectively opened the doors of many banks twenty-four hours a day anyway. Unlike bank employees, ATMs don't complain when they aren't excused to leave for the golf course at 3 P.M.

Submitted by Dorio Barbieri of Mountain View, California.
Thanks also to Herbert Kraut of Forest Hills, New York.

WHEN DO FISH SLEEP?   1

Speaking of ATMs . . . When They Were Introduced, ATMs Were Supposed to Save Labor Costs for the Banks and Ultimately Save Money for the Customers. Now My Bank Is Charging Money for Each ATM Transaction. What Gives?

The banking industry is being squeezed from two sides. On the one hand, customers now demand interest on checking accounts and money-market rates on savings accounts. Yet they also want services provided for free.

While it is true that an ATM transaction generally is cheaper for the banks than the same transaction conducted by a teller, banks have spent a fortune buying and installing these ma-chines. As David Taylor, of the Bank Administration Institute ⌐ it, "As the customer gets more and more convenience and ᵗrol of his banking options, he will have to pay for each op-one at a time." The alternative would be a return to having ʳvice fees but also to customers getting lower interest rates

DAVID FELDMAN

on CDs and checking and savings accounts, which banks know would be suicidal for them. As bank deregulation accelerates and banks are allowed to compete with brokerages and other financial institutions, expect to see increasing service charges.

Most banks do not charge for ATM transactions. If there are two big banks in a town, each knows that if it charges for ATM transactions, the other bank will advertise that its machines are free. So the choice between free and pay ATMs is left to what the banking business calls "competitive reasons," which is fiduciary lingo for "if we think we can get away with charging for it, we will."

Why Does Granulated Sugar Tend to Clump Together?

It ain't the heat, it's the humidity. Sugar is hygroscopic, meaning that it is capable of absorbing moisture from the air and changing its form as a result of the absorption. When sugar is subjected to 80% or higher relative humidity, the moisture dissolves a thin film of sugar on the surface of the sugar crystal. Each of these crystals turns into a sugar solution, linked to one another by a "liquid bridge."

According to Jerry Hageney, of the Amstar Corporation, when the relative humidity decreases, "the sugar solution gives up its moisture, causing the sugar to become a crystal again. The crystals joined by the liquid bridge become as one crystal. Thus, hundreds of thousands of crystals become linked together to form a rather solid lump."

Although we can't see the moist film on sugar exposed to high humidity, it won't pour quite as smoothly as sugar that has never been exposed to moisture. But when it dries up again, the liquid bridge is a strong one. Bruce Foster, of Sugar Industry Technologists, told us that the technology used to make sugar cubes utilizes this natural phenomenon.

To make sugar cubes, water is added to sugar in a cube-shaped mold. After the sugar forms into cubes, it is dried out, and voilà! you have a chemical-free way to keep sugar stuck together.

Submitted by Patty Payne of Seattle, Washington.

Why Do Two Horses in an Open Field Always Seem to Stand Head to Tail?

Horses, unlike people, don't bother to make the pretense of listening to what companions have to say. And also, unlike humans, horses have tails. Rather than stand around face-to-face boring each other, figures the horse, wouldn't it be more practical to stand head to tail? This way, with one swish of the tail, a horse can rid its body of flies and other insects while knocking the bugs off of the head of the other horse.

In cold weather, horses are more likely to stand head-to-head, so they can help keep each other warm with their breaths. In this one respect, horses are like people—they are full of an inexhaustible supply of hot air.

Submitted by Mrs. Phyllis A. Diamond of Cherry Valley, California.

Why Does Your Whole Body Ache When You Get a Cold or Flu?

When a virus enters your bloodstream, it releases several compounds that mount your body's defense against infection. Interferon, interleukin, and prostaglandins are among the body's most

DAVID FELDMAN

valuable compounds. They raise a fever, shift the metabolism, and increase blood flow to areas of the body that need it.

Frank Davidoff, of the American College of Physicians, suggests that although science hasn't yet precisely defined their function, there is much evidence to suggest that these compounds are responsible for the aching feeling that accompanies colds and flus. More of the compounds are usually found in the bloodstream during the aching phase than before any symptoms start. And when doctors inject a purified form of each compound into a patient, many of the symptoms of a virus, including fever, sweating, and aching, occur without actually causing the entire illness.

These compounds are effective without anyone knowing precisely *how* they work, but there are logical explanations for *why* they work. Davidoff sums it up well:

> the aching and other symptoms seem to be the "price" that's paid for mounting a defense against the infection. Whether the price is inseparable from the defense isn't clear. Thus, on the one hand, the symptoms might actually be a holdover from some mechanism that was important earlier in evolution but that is unnecessary now in more complex creatures. On the other hand, symptoms like aching may be part and parcel of the defense; I don't believe anyone knows for sure.

Submitted by James Wheaton of Plattsburgh Air Force Base, New York.

HOW Did Romans Do the Calculations Necessary for Construction and Other Purposes Using Roman Numerals?

Our idea of a good time does not include trying to do long division with Roman numerals. Can you imagine dividing CXVII by IX and carrying down numbers that look more like a cryptogram than an arithmetic problem?

The Romans were saved that torture. The Romans relied on the Chinese abacus, with pebbles as counters, to perform their calculations. In fact, Barry Fells, of the Epigraphic Society, informs us that these mathematical operations were performed in Roman times by persons called "calculatores." They were so named because they used *calcule* (Latin for pebbles) to add, subtract, multiply, and divide.

Submitted by Greg Cox of San Rafael, California.

Why Do Some Ice Cubes Come Out Cloudy and Others Come Out Clear?

A caller on the Merle Pollis radio show, in Cleveland, Ohio, first confronted us with this problem. We admitted we weren't sure about the answer, but subsequent callers all had strong convictions about the matter. The only problem was that they all had *different* convictions.

One caller insisted that the mineral content of the water determined the opacity of the cube, but this theory doesn't explain why all the cubes from the same water source don't come out either cloudy or clear.

Two callers insisted that the temperature of the water when put into the freezer was the critical factor. Unfortunately, they couldn't agree about whether it was the hot water or the cold water that yielded clear ice.

We finally decided to go to an expert who confirmed what we expected—all the callers were wrong. Dr. John Hallet, of the Atmospheric Ice Laboratory of the Desert Research Institute in Reno, Nevada, informed us that the key factor in cloud formation is the temperature of the *freezer*.

When ice forms slowly, it tends to freeze first at one edge. Air bubbles found in a solution in the water have time to rise and escape. The result is clear ice cubes.

DAVID FELDMAN

The clouds in ice cubes are the result of air bubbles formed as ice is freezing. When water freezes rapidly, freezing starts at more than one end, and water residuals are trapped in the middle of the cube, preventing bubble loss. The trapped bubbles make the cube appear cloudy.

Why Are Most Pencils Painted Yellow?

Pencils came in various colors before 1890, but it was in that year the Austrian L & C Hardtmuth Company developed a drawing pencil that was painted yellow. Available in a range of degrees of hardness, the company dubbed their product Koh-I-Noor.

In 1893, L & C Hardtmuth introduced their Koh-I-Noor at the Chicago World's Colombian Exposition, and Americans responded favorably. Ever since, yellow has been synonymous with quality pencils.

Monika Reed, product manager at Berol USA, told *Imponderables* that although Berol and other manufacturers make pencils painted in a wide range of colors, yellow retains its great appeal. According to Bill MacMillan, executive vice president of the Pencil Makers Association, sales of yellow-painted pencils represent 75% of total sales in the United States.

Submitted by Robert M. Helfrich of Pittsburgh, Pennsylvania.
Thanks also to Beth Newman of Walnut Creek, California.

DAVID FELDMAN

Why Do You Have to Use #2 Pencils on Standardized Tests? What Happens If You Use a #1 Pencil? What *Is* a #2 Pencil?

If only we could blame our SAT scores on using #1 or #3 pencils! But it's hard to find any other besides #2s anyway.

All-purpose pencils are manufactured in numbers one through four (with half sizes in between). The higher the number, the harder the pencil is. Although the numbers of pencils are not completely standardized, there is only slight variation among competitors.

The #2 pencil, by far the most popular all-purpose pencil, is considered medium soft (compared to the #1, which is soft; to #2.5, medium; to #3, medium hard; and to #4, hard). Pencils are made harder by increasing the clay content and made softer by increasing the graphite content of the lead.

Why do some administrators of standardized tests insist on #2 pencils? Because the degree of hardness is a happy compromise between more extreme alternatives. A hard pencil leaves marks that are often too light or too thin to register easily on mark-sensing machines. Too soft pencils, while leaving a dark mark, have a tendency to smudge and thus run into the spaces left for other answers.

Even some #2 pencils might not register easily on mark-sensing machines. For this reason, Berol has developed the Electronic® Scorer. According to Product Manager Monika Reed, "This pencil contains a special soft lead of high electric conductivity," which eases the burden of today's high-speed marking machines.

Unfortunately, even the Electronic Scorer doesn't come with a guarantee of high marks, only accurately scored answers.

Submitted by Liz Stone of Mamaroneck, New York. Thanks also to John J. Clark of Pittsburgh, Pennsylvania; Gail Lee of Los Angeles, California; William Lush of Stamford, Connecticut; and Jenny Bixler of Hanover, Pennsylvania.

Why Do Fish Eat Earthworms? Do They Crave Worms or Will Fish Eat Anything that Is Thrust upon Them?

We have to admit, earthworms wouldn't be our first dining choice. What do fish see in worms that we don't see (or taste)?

R. Bruce Gebhardt, of the North American Native Fishes Association, emphasizes that just about any bait can entice a fish if the presentation is proper. Human gourmets may prefer a colorful still life on white china, but fish prefer a moving target. And they are a little less finicky than humans:

> A pickerel, for example, will attack a lure before it's hit the water. It must instantly assess the size of the bait; if it's a pine cone, it will worry about spitting it out after it is caught.

Most fish are attracted to food by sight, and prefer live bait. Fish are often attacking and testing as much as dining:

> It is unnecessary to *completely* convince the fish that the bait is alive. Most fish encountering anything strange will mouth it or closely examine it as potential food; the less opportunity it's liable to have, the more vigorously it will attack.

While the fisherman might think that every pull on his line means the fish finds his worm irresistible, the fish may well be nibbling the worm to determine the identity of the bait—by the time it finds out it has caught a worm, it's too late: It is hooked.

Our Imponderable also assumes that fish may go out of their way to eat earthworms, but Gerry Carr, director of Species Research at the International Game Fish Association, assures us that given a choice, most fish will go after food native to their environment:

> Nature is constituted in a way that everything has its place and is in ecological balance. Fish eat the foods that nature provides for them. The fly fisherman is acutely aware of this. He or she knows that trout, for example, at a certain time of year, seem to crave and feast on the type of nymphs that are hatching and falling into the water at that moment. Any other kind of artificial fly will not work, only the one that best imitates the hatch.
>
> Of course, not all fish are that finicky. Catfish eat anything that stinks. Logical! Their purpose in nature is to clean up the bottom, eliminating dead, rotting carcasses that rob water of oxygen and might cause all the fish to die. Nature's vacuum cleaner! And they survive because they have carved out or been given an ecological niche in the system that is not overly in competition with other species.

But why will worms attract even finicky fish? Carr continues:

> Worms, actually, are probably more of a side-dish in the diets of some fishes, a sort of aperitif. Worms look tasty, so the fish eats them. I do not think fish go looking for worms, specifically, unless they have got their appetite whet up for them by an angler conveniently drowning them.

Even if worms aren't native to a fish's environment, they fulfill most of the prerequisites for a favorite fish fast food. The size and shape are good for eating, and the fact that worms are wiggling when alive or look like they are moving even when dead adds to their allure. Carr mentions that barracudas cannot resist any appropriately sized bait or lure that is long and slender or cigar-shaped and moving at the right speed. "But offering them a worm that just sits there would be tantamount to a human asking for jelly instead of 'All-Fruit.'"

WHEN DO FISH SLEEP?

One other point needs to be stated. The popularity of earthworms as bait is undoubtedly enhanced by the cheapness, easy availability, and convenience of them. As Gebhardt put it, "It's probably anglers' convenience that has given earthworms their reputation for delectability rather than petitions signed by fish."

Submitted by Roy Tucker of Budd Lake, New Jersey.

Why Are Stock Prices Generally Quoted in Eighths?

In *Why Do Clocks Run Clockwise? and Other Imponderables* we discussed the derivation of our expression "two bits." In Spain, a *bit* was one of the "pieces of eight," an actual pie-shaped slice of a peso. Two bits were one-quarter of a peso.

Spanish coins circulated freely in the New World before and during colonial times for at least two reasons. There weren't enough native coins to go around, and Spanish gold and silver specie were negotiable just about anywhere in the world (like in the good old days when foreign nations sought American dollars) because they were backed by gold.

Was it a coincidence that two bits of a peso happened to equal two bits of a dollar? Not at all. Peter Eisenstadt, research associate at the New York Stock Exchange archives, told *Imponderables* that when U.S. currency was decimalized in 1785,

> the U.S. silver dollar was established with a value equivalent to the Spanish silver peso. Though the official divisions of the dollar were in decimals, many continued to divide the new U.S. dollar into eighths and this practice was followed in securities trading.

Stocks were usually traded in eighths from the inception of securities trading in the United States in the 1790s. Eisenstadt believes that Americans simply borrowed the practice of quoting in eighths from the Europeans. As he notes, most early stockbrokers were part-timers, devoting most of their attention to the

DAVID FELDMAN

merchant trade, which had long quoted prices in eighths.

By the 1820s, stocks traded on the NYSE were universally quoted in eighths, but this was an informal arrangement; it became a requirement in 1885. The American and Pacific Stock Exchanges followed suit.

Although the history of our quoting stock prices in eighths makes historical sense, we don't understand why the exchanges still maintain the practice. When a stock dips to near zero, prices now are quoted in sixteenths and even thirty-seconds of a dollar, forcing financial tycoons to rely on their memory of grade-school fractional tables when doing calculations. And what happens when someone wants to sell his one share of stock quoted at 48⅜? Who gets the extra half-cent?

Wouldn't it make more sense to quote all stocks in hundredths of a dollar? Why should two-dollar stocks have to rise or fall more than ten percent at a time when a 2% change in most stocks is considered significant? Roy Berces, of the Pacific Stock Exchange, acknowledges that our system is probably archaic, but sees no groundswell for changing tradition.

Submitted by E. B. Peschke of St. Charles, Missouri. Thanks also to John A. Bush of St. Louis, Missouri; Christopher Dondlinger of Longmont, Colorado; and Dave Klingensmith of Canal Fulton, Ohio.

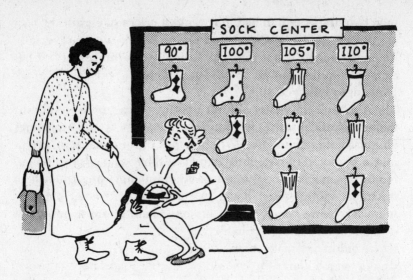

Why Are Socks Angled at Approximately 115 to 125 Degrees When the Human Foot Is Angled at About 90 Degrees?

Not all socks are angled, of course. Tube socks are "angled" at 180 degrees. Tube socks are so named because they are a straight tube of fabric closed on one end by sewing. The tube sock is constructed by "full circular knitting" (i.e., the knitting head on the machine knits in a full circle).

A tube sock doesn't contain a designated position for the heel, but more conventional socks do. Most socks are knitted with a feature called the "reciprocated heel." Sid Smith, president and chief executive officer of the National Association of Hosiery Manufacturers, told *Imponderables* how the reciprocated heel is made:

> Imagine a full circular knitting machine starting at the top of the sock and knitting in a complete circle all the way down the top of the sock, until it hits the point where the heel is to be knitted in. At this point, the machine automatically enters what is called the "reciprocated function." Instead of knitting in a complete circle,

DAVID FELDMAN

it knits halfway to each side and then back again, until the heel portion is knitted in.

After this is completed, the machine automatically reverts to full circular knitting to finish the sock. This reciprocation is what causes the finished sock to be angled.

The 115- to 125-degree angle of the sock, then, is the result of, rather than the purpose of, the knitting process. The fabrics used for socks will give or stretch to conform to the contours of the foot. Since a 180-degree tube sock can fit comfortably on the human foot, there is no reason why a conventional sock won't.

Submitted by Vernon K. Hurd of Colorado Springs, Colorado.

Why Do Cattle Guards Work?

No, there aren't demons underground shooting BB pellets between the bars of the cattle guards. Cows are afraid to walk where their feet can't get solid footing.

Our correspondent mentions that he has seen painted white strips used as cattle guards, presumably tricking cows into thinking that the unpainted area is a black hole. Cows are evidently as subject to phobias as cowboys and cowgirls.

Submitted by A. M. Rizzi of Torrey, Vermont.

Why Are There No A- or B-Sized Batteries?

Because they are obsolete. A- and B-sized batteries once existed as component cells within much larger zinc carbon battery packs. The A cells supplied the low-voltage supply for the filaments in the vacuum tubes used to supply power to early radios and crank telephones.

Of course, the descendants of the old A- and B-sized batteries are still with us. As electronic devices have gotten smaller, so have the batteries that power them. As might be expected, the A cell came first, then B, C, and D cells. The batteries were lettered in ascending order of size. James Donahue, Jr., of Duracell, Inc., says that as cells smaller than the original A cells were developed, they were designated as AA and then AAA cells. Donahue reports that there is even a new AAAA battery.

So the old A- and B-sized batteries are no longer in production. It's no use having a battery larger than the device it powers.

Submitted by Larry Prussin of Yosemite, California. And thanks also to Herman E. London of Poughkeepsie, New York; Nancy Ondris of Kings Park, New York; and Ronald Herman of Montreal, Quebec.

DAVID FELDMAN

What Are Those Little Plastic Circles (that Sometimes Have Rubber in the Middle) Found on the Walls of Hotels?

If you've noticed, those circles are located about three feet off the ground and usually near the entrance. They are called wall protectors, and their sole function in life is to keep doorknobs from slamming against the walls. And with some of the paper-thin walls we've encountered in motels, wall protectors may be responsible for keeping the structural integrity of the building intact.

Submitted by Carol Rostad of New York, New York.

Why Does Starch Make Our Shirts Stiff?

Starch is a type of "sizing," a filler used to add body, sheen, and luster to limp clothing. All shirts come off the rack with sizing, but sizing is water-soluble; every time the shirt is washed, sizing comes out of the shirt. The main purpose of adding starch, then, is to restore the original body of a garment.

The main ingredient in starch is wheat or, less frequently, corn. The grain is mixed with water, resins, and chemicals. As Bill Seitz, of the Neighborhood Cleaners Association, describes it, the starch is literally absorbed by the fabric. Cotton plus wheat is stiffer than cotton alone.

Norman Oehlke, of International Fabricare Institute, adds that starch also enhances soil resistance, facilitates soil removal for the next wash, and makes ironing easier.

Synthetic fabrics aren't as receptive to starch as all-cotton garments, so extra chemicals are added to the starch, such as polyvinyl acetate, sulfated fatty alcohols, silicones, and our personal favorite, carboxymethylcellulose.

Submitted by Kris Heim of De Pere, Wisconsin. Thanks also to Stanley R. Sieger of Pasadena, California.

How Does the Campbell Soup Company Determine Which Letters to Put in Their Alphabet Soup? Are There an Equal Number of Each Letter? Or Are the Letters Randomly Inserted in the Can?

We spoke to a delightful young woman at Campbell's named Ginny Marcin, who, astonishingly, did not have the answers to these questions at her fingertips. But she spoke to the vice president of Letter Distribution and obtained the following information.

Campbell's makes two sizes of letters for their soups. Small letters go into some of the prepared soups (such as the Chunky line). Slightly larger letters bejewel their vegetable and vegetarian vegetable soups.

It is the stated intention of the Campbell Soup Company not to discriminate against any letter. All are equally represented. However, Campbell's cannot control the distribution of letters while inserting the letters and soup into the can, so irregularities can result. You might find a can with eight q's and only three u's, screwing up your plans to use the letters as Scrabble tiles.

Come to think of it, if the letters really are distributed randomly, why does Campbell's need a vice president of Letter Distribution?

Submitted by Tom Carroll of Binghamton, New York.

What Is the Purpose of Corn Silk?

These strands, which bedevil shuckers and flossless eaters alike, actually do have an important purpose. The longer threads of corn silk stand outside of the husk in tufts to collect pollen. The pollen then travels the silk to the ear of corn and fertilizes it.

Edith M. Munro, director of Information of the Corn Refiners Association, told *Imponderables* one of the critical factors exacerbating the loss in the corn harvest during the 1988 drought was that the "lack of moisture delayed the development of silks or dried the silks up, so that no silks were present when pollen was released." Without sufficient pollination, the growth of the corn is stunted, resulting in ears of corn with only a few kernels.

Submitted by Denise Dennis of Shippensburg, Pennsylvania.

DAVID FELDMAN

Why Are U.S. Government Department Heads Called "Secretaries" Instead of "Ministers," as in Most Other Countries?

The word "secretary" comes from the same Latin root as the word "secret." In medieval days, a secretary was a notary or a scribe, someone privy to secret and often important information. Over time, secretaries became not only men and women in charge of correspondence for an employer but trusted advisors to heads of state and royalty. So although today's office secretaries may now be a neglected and abused lot, Europeans have long called important officeholders "secretaries."

We wrote to several historians who were kind enough to unravel this Imponderable. They concurred that although Americans appropriated their governmental vocabulary from the English, no single term was used to describe cabinet-level officials in England at the time the United States Constitution was drafted.

The parliamentary-cabinet style government of England was not established until the early 1700s, and many of the titles from feudal governments still existed. Thomas L. Purvis, of the Institute of Early American History and Culture, elaborates on the mishmash of English titles:

> members of the cabinet carried titles both feudal and modern, such as Chancellor of the Exchequer and Prime Minister. Intermediate in age were the secretaries of various departments, such as the former Secretary of State for the Southern Department (whose purview extended over the American Colonies) and the ad hoc Secretary *at* War.

Americans, in their revolutionary ethos, were not about to give a nod to the hated English king and his ministers. The terms "president" and "vice president" were chosen to distinguish elected leaders from the dreaded monarchy.

None of the framers of the Articles of Confederation wrote why "secretary" was designated as the term for America's executive officers. The Department of the Treasury conducted an investigation into this Imponderable and found that the Library of Congress, the National Archives, and the Office of Protocol at the Department of State could provide no documentary evidence for the choice.

But all of our sources indicated that the attempt to distance the United States from any trappings of a monarchy contributed to the selection of "secretary." Samuel R. Gammon, executive director of the American Historical Association, told *Imponderables* that "the older English tradition of terming the monarch's chief executive assistants 'Principal Secretary of State' may also have been in their [the framers of the Constitution] minds."

"Secretary" was a solid, middle-of-the-road choice. As Purvis points out, the title seems honorific yet confers no indication of aristocracy and could be applied to any department in the government.

Submitted by Daniel Marcus of Watertown, Massachusetts.

Why Is Prepackaged Chocolate Milk Thicker in Consistency than the Chocolate Milk You Make at Home?

Gravity.

If you make a batch of chocolate milk at home and put it in the refrigerator to cool, you will notice something when you fetch it ten hours later. The chocolate sinks to the bottom.

All is not lost. Simply shaking up the container will redistribute the chocolate throughout the milk.

But this kind of separation is unacceptable in a commercial product, especially one that is sold in a transparent container. So commercial dairies use stabilizers and emulsifiers to assure that the chocolate and milk remain mixed. Although the job of the (usually natural) stabilizers and emulsifiers is to keep the choc-

DAVID FELDMAN

olate from falling to the bottom of the carton, the by-product is a thicker consistency than home-style chocolate milk.

Submitted by Herbert Kraut of Forest Hills, New York.

Why Do Fingernails Grow Faster than Toenails?

This is not the kind of question whose solution wins Nobel Prizes for scientists or garners prestigious grants for research hospitals, yet the answer is not obvious. The average severed fingernail takes four to six months to grow back to its normal length. The average toenail takes nine to twelve months.

Dermatologist Dr. Fred Feldman says that although nobody knows for sure why toenails lag behind fingernails in growth, there are many possible explanations:

1. Trauma makes nails grow faster. Dermatologists have found that if a patient bites a nail down or loses it altogether, the traumatized nail will grow faster than on one left alone. Fingernails, in constant contact with many hard or sharp objects, are much more likely to be traumatized in everyday life than toenails. Even nonpainful contact can cause some trauma to nails. Because we use our fingers much more often than our toes, toenails do not tend to get the stimulation that fingernails do.
2. All nails grow faster in the summer than the winter, which suggests that the sun promotes nail growth. Even during the summer, most people cover their toenails with socks and shoes.
3. Circulation is much more sluggish in the feet than in the hands.

Our medical consultants did not suggest the obvious: The faster growth of fingernails is nature's way of providing us with a constant tool with which to open pistachio nuts.

Submitted by Dave Bohnhoff of Madison, Wisconsin.

WHEN DO FISH SLEEP?

Why Do We Dream More Profusely When We Nap than We Do Overnight?

According to the experts we consulted, we dream just as much at night as we do when we take a nap. However, we *recall* our afternoon-nap dreams much more easily than our dreams at night.

While we are dreaming, our long-term memory faculties are suppressed. During the night, our sleep is likely to go undisturbed. We tend to forget dreams we experience in the early stages of sleep. The sooner that we wake up after having our dreams, the more likely we are to remember them.

Any situation that wakes us up just after or during the course of a dream will make the sleeper perceive that he or she has been dreaming profusely. Dr. Robert W. McCarley, the executive secretary of the Sleep Research Society, told *Imponderables* that women in advanced stages of pregnancy often report that they are dreaming more frequently. Dr. McCarley believes that the perceived increase in dreaming activity of pregnant women is prompted not by psychological factors but because their sleep is constantly interrupted by physical discomforts.

Why Do Place Kickers and Field-Goal Kickers Get Yardage Credit from Where the Ball Is Kicked and Yet Punters Only Get Credit from the Line of Scrimmage?

Well, who said life was fair? It turns out that this blatant discrimination occurs not because anyone wants to persecute punters particularly but for the convenience and accuracy of the scorekeepers. Jim Heffernan, director of Public Relations for the National Football League, explains:

DAVID FELDMAN

Punts are measured from the line of scrimmage, which is a defined point, and it sometimes is difficult to determine exactly where the punter contacts the ball. Field goals are measured from the point of the kick because that is the defined spot of contact.

Submitted by Dale A. Dimas of Cupertino, California.

HOW Does a Gas Pump "Know" When to Shut Off When the Fuel Tank Is Full?

A sensing device, located about one inch from the end of the nozzle, does nothing while fuel is flowing into the gas tank, but is tripped as soon as fuel backs up into the nozzle. The sensing device tells the nozzle to shut off.

Because of the location of the sensing device and the relatively deep position of the nozzle, a gas tank is never totally filled unless the customer or attendant "tops off" the tank. Topping off tanks is now illegal in most states and is a dangerous practice anywhere.

Submitted by Stephen O. Addison, Jr. of Charlotte, North Carolina.

HOW Does the Treasury Know When to Print New Bills or Mint New Coins? How Does it Calculate How Much Money Is Lost or Destroyed by the Public?

There are more than two hundred billion dollars in coins and currency in circulation today in the United States. Determining the necessary timing for the minting and printing of new monies is therefore far from a simple task.

Most of the demand for new money comes from banks. When a bank receives more checks to cash than it can comfortably accommodate with its cash on hand, the bank orders new money from one of the twelve Federal Reserve Banks. Of course, the bank doesn't get the new money for free; it uses a special checkbook to order new cash. When a bank has excess cash, it can deposit money into an account at the Federal Reserve Bank to offset its withdrawals.

What happens when the Federal Reserve Bank itself runs out of coins or notes? It places an order with the U.S. Mint for new coins or the Bureau of Engraving and Printing for the new

DAVID FELDMAN

currency. So demand from individual banks, funneled through a larger "distributor"—a Federal Reserve Bank—is responsible for the decision to issue new currency.

The average life-span of a dollar bill is fifteen to eighteen months. Larger denominations tend to have a longer life because they are circulated less frequently. The perishability of paper notes is the second major factor in calculating the requirements for new currency. In 1983 alone, the twelve Federal Reserve Banks destroyed more than 4.4 billion notes, worth more than $36 billion. The constant retirement of defective bills explains why almost one out of every four notes the Federal Reserve Bank sends to local banks is a newly printed one.

Every time a Federal Reserve Bank receives currency from a local bank, it runs the notes through high-speed machines designed to detect unfit currency. The newest machines can inspect up to sixty thousand notes per hour, checking each bill for dirt by testing light reflectivity (the dirtier the note, the less light is reflected) and authenticity (each note is tested for magnetic qualities that are difficult for counterfeiters to duplicate).

Notes valued at $100 or less are destroyed by the local Federal Reserve Bank. Unfit bills used to be burned and processed into mulch (we kid you not), but they are now shredded and compressed into four-hundred-pound bales. Most of these bundles of booty are discarded at landfills. Federal Reserve notes in denominations of $500 or more are canceled with distinctive perforations and cut in half lengthwise. The local Federal Reserve Bank keeps the upper half of each note and sends the other half to the Department of Treasury in Washington, D.C. When the Treasury Department verifies the legitimacy of the notes, it destroys its halves and informs the district bank that it may destroy the upper halves.

Coins have a much longer life in circulation, but the Mint still produces more than 50 million coins a day (compared to "only" twenty million notes printed per day). A U.S. Mint official told us that shipping coins across country is not a trivial task logistically—five-hundred-thousand pennies, for example, are a tad bulky. Huge tractor-trailer trucks, up to 55 feet in length and

13½ feet high, are used to transport coins from the Mint to Federal Reserve Banks. Dimes, quarters, and half dollars are transported by armored carriers.

The demand process for coins works the same way as for paper notes. Although the Mint has learned that seasonal peaks run true from year to year (the demand for coins goes up during prime shopping seasons, such as Christmas), the Mint yields to the demands of its constituent Federal Reserve Banks.

Submitted by Hugo Kahn of New York, New York.

DAVID FELDMAN

What Is the Purpose of that Piece of Skin Hanging from the Back of Our Throat?

No, Kassie Schwan's illustration to the contrary, the purpose of that "hanging piece of skin" is not to present targets for cartoon characters caught inside other characters' throats. Actually, that isn't skin hanging down, it's mucous membrane and muscle. And it has a name: the uvula.

The uvula is a sort of anatomical tollgate between the throat and the pharynx, the first part of the digestive tract. The uvula has a small but important role in controlling the inflow and out-flow of food through the digestive system. Dr. William P. Jollie, chairman, Department of Anatomy, the Medical College of Virginia, explains: "The muscle of both the soft palate and the uvula elevates the roof of the mouth during swallowing so that food and liquid can pass from the mouth cavity into the pharynx."

Dr. L.J.A. DiDio, of the Medical College of Ohio, adds that the uvula also helps prevent us from regurgitating our food during swallowing. Without the uvula, some of our food might enter the nasal cavity, with unpleasant consequences.

Submitted by Andy Garruto of Kinnelon, New Jersey.

WHEN DO FISH SLEEP?

Why Don't Birds Tip Over When They Sleep on a Telephone Wire?

A telephone wire, of course, is only a high-tech substitute for a tree branch. Most birds perch in trees and sleep without fear of falling even during extremely windy conditions.

The secret to birds' built-in security system is their specialized tendons that control their toes. The tendons are located in front of the knee joint and behind the ankle joint. As it sits on its perch, the bird's weight stretches the tendons so that the toes flex, move forward, and lock around the perch.

Other tendons, located under the toe bones, guarantee that a sleeping bird doesn't accidentally tip over. On the bottom of each tendon are hundreds of little projections. These fit perfectly into other ratchetlike sheaths. The body weight of the bird pressing against the telephone wire (or tree branch) guarantees that the projections will stay tightly locked within the sheaths.

Barbara Linton, of the National Audubon Society, adds that while this mechanism is most highly developed in perching

DAVID FELDMAN

birds and songbirds, many other birds do not perch to sleep. They snooze on the ground or while floating on water.

Submitted by Dr. Lou Hardy of Salem, Oregon. Thanks also to Jann Mitchell of Portland, Oregon.

Why Is It Sometimes Necessary to Stroke a Fluorescent Lamp to Get It to Light?

All fluorescent bulbs require a ground plane to start. If the fluorescent lamp is inside a metal fixture, any piece of metal, such as the reflector, can serve as a ground plane. Richard H. Dowhan, manager of Public Affairs for GTE Products Corporation, told *Imponderables* that the closer the ground plane is to the tube, the easier it is to start the fluorescent. "Placing your hand on the tube or stroking it creates a very effective ground plane." Magicians have been lighting "naked" fluorescent bulbs for quite a long time by serving as the ground plane.

But most of us aren't magicians, and most of us use fluorescent lamps inside of metal fixtures. Why do the lamps usually light with a flick of the switch at some times and then other times require a little massage? J. Robert Moody, of General Electric's Lighting Information Center, was kind enough to supply an answer that doesn't require a physics degree to understand.

Under normal conditions, fluorescent lamps should light without difficulty, with the electric current flowing inside the fluorescent tube. But if the lamp has a combination of a light coating of dust and a small amount of moisture from the air, the coating will allow "some of the electric current to flow on the outside of the tube, and the current on the outside of the bulb will prevent the lamp from lighting. Under this condition, stroking the tube will interrupt the flow of current on the outside of the tube and cause the light to come on."

Submitted by Harold J. Ballatin of Palos Verdes, California.

Why Is There an Expiration Date on Sour Cream? What's the Matter, Is It Going to Get More Sour?

We've gotten this Imponderable quite often on radio interviews, usually from smug callers sure that expiration dates are a capitalist plot to force us to throw away barely used sour cream. But mark our words: if you think sour cream is tart when you open it, just leave it in the refrigerator too long and taste the difference. As the expiration date on sour cream becomes a dim memory, bacteria acts upon the sour cream, making it unbearably tart. Given enough time, mold will form on the sour cream, even if it is properly refrigerated.

Sour cream has about a month-long life in the refrigerator. Wait much longer and we'll bet that you won't want to test just how sour cream can get. If you think we're wrong, there's one way to find out for sure.

Go ahead and taste it. Make our day.

DAVID FELDMAN

Who Translates the Mail When a Letter Is Sent to the United States from a Foreign Country that Uses a Different Alphabet?

If the United States Postal Service has problems sending a letter across town in a few days, we wondered how they contended with a letter sent to Nebraska from a remote village in Egypt. Does every post office hire a staff of linguists to pore over mail and route it in the right direction?

No, not every post office. But the USPS does employ linguists at their International Exchange Offices, located at the major ports (New York, San Francisco, Miami, and Boston) where foreign mail is received. All mail is separated and sorted at these border points and sent on its merry way.

We contacted some foreign consulates to find out how they solved the problem of indecipherable mail. A representative of the Greek consulate told *Imponderables* that if foreign mail is written in one of the international languages, multilingual personnel have no problem sorting it. If no postal worker can translate an address, the postal service will likely do what we did— call the embassy or consulate of the country of the sender and hope for the best.

Submitted by Charles F. Myers of Los Altos, California.

Why Do Roaches Always Die on Their Backs?

We couldn't believe that three readers actually had experienced the good fortune to see a dead roach and had torn themselves away from the subsequent celebration long enough to note the posture of the deceased insect. But we trudged on nevertheless, contacting entomologists who actually get paid to study stuff like this.

Professor Mary H. Ross, affiliated with Virginia Polytechnic Institute and State University, told *Imponderables* that when a roach dies, its legs stiffen and the cockroach falls on its side. Because most roaches have a flattened body form with narrow sides, the momentum of the fall rolls them onto their backs.

John J. Suarez, technical manager of the National Pest Control Association, adds that small cockroaches, such as the German and the brown-banded, are more likely to die on their backs. Larger cockroaches with lower centers of gravity, such as the American and the Oriental, occasionally die face down.

Needless to say, we can't guarantee the position of dead roaches contained in traps. Maybe the lifeless occupants of Roach Motels lie perfectly prone. Unfortunately, there is only one way to find out and only entomologists have the stomach for it. Please don't try to verify this at home!

Submitted by Gloria Stiefel of Orange Park, Florida. Thanks also to Irma Keat of Somers, New York; and Gregg Hoover of Morgan Hill, California.

Why Does Warmth Alleviate Pain?

A caller on Tom Snyder's radio show posed this Imponderable. We had no idea of the answer, but it was surprising that so many physicians we spoke to didn't know the answer either.

We finally got the solution from Daniel N. Hooker, Ph.D., coordinator of Physical Therapy/Athletic Training at the University of North Carolina at Chapel Hill. His answer included plenty of expressions like "receptors," "external stimuli," and "pain sensors." So let's use an analogy to simplify Hooker's explanation.

If a pneumatic drill is making a ruckus outside your window, you have a few choices. One is to do nothing, which won't accomplish much until the drill stops. But another option is to go to your stereo and put on a Led Zeppelin record at full blast. The pneumatic drill is still just as loud—you may still even be able to hear it. But the music will certainly distract you (and for that matter, your next-door neighbors as well), so the drilling doesn't seem as loud.

Hooker emphasizes that most of us associate warmth with pleasant experiences from our youth. By placing heat on the part of our body that hurts we stimulate the sensory receptors, which tell our brain that there has been a temperature change. This doesn't eliminate the pain, but the distraction makes us less aware of the pain. As our body accommodates to the high temperature, we need fresh doses of warmth to dampen the pain. When we receive the renewed heat treatment, we *expect* to feel better, so we do.

Why Can't We Use Both Sides of a Videotape like We Do with an Audio Tape?

Don French, chief engineer of Radio Shack, is getting a little testy with us: "If you keep using me as a consultant on your books, we are going to have to start charging for my service!"

We have read all of the bestselling business management books. They all reiterate that most people aren't motivated by higher pay but by recognition of their effort and accomplishments. So to you, Don French, we want to acknowledge our heartfelt appreciation for the efforts you have expended in educating the American public on the wonders and intricacies of modern technology in our contemporary culture of today. Through your efforts, our citizens will be better equipped to handle the challenges and complexities of the future.

But not one penny, bub.

Luckily, Mr. French couldn't resist answering this Imponderable anyway.

DAVID FELDMAN

It turns out that even though some audio cassette recorders require the tape to be flipped before recording on the other side, the recorder doesn't actually copy on both sides of the tape. It copies on the top side of the tape in one direction and the bottom in the other direction.

On videotapes, the audio is also recorded on a small portion of the top side of the tape. But the video, with a much higher frequency requirement and slower recording speed, needs much more room to copy, and is recorded diagonally on most of the remaining blank tape.

Submitted by Jae Hoon Chung of Demarest, New Jersey.

Why Are the Toilet Seats in Public Restrooms Usually Split Open in the Front?

This has become one of our most frequently asked Imponderables on radio shows. So for the sake of science and to allay the anxiety of unspoken millions, here's the, pardon the expression, poop on a mystery whose answer we thought was obvious.

Try as they might, even the most conscientious janitors and bathroom attendants know it is impossible to keep a multiuser public toilet stall in topnotch sanitary condition. Let's face it. Pigs could probably win a slander suit from humans for our comparing our bathroom manners to theirs. Too many people leave traces of urine on top of toilet seats. Men, because of a rather important physiological distinction from women, particularly tend not to be ideally hygienic urinators, but most sanitary codes make it mandatory that both male and female toilets contain "open-front" toilet seats in public restrooms. In fact, at one time, "open back" seats were mandated as well, but the public wouldn't stand (or sit) for them.

If they are more hygienic, why not use open-front toilet seats at home? The answer is psychological rather than practical. An open-front seat would imply to the world that one's bathroom habits were as crass as those employed by the riffraff who use public restrooms. Still, we would think that open-front toilet seats in home bathrooms might lessen the number of divorce-causing arguments about men keeping the toilet seats up.

Submitted by Janet and James Bennett of Golden, Colorado. Thanks also to Tom Emig of St. Charles, Missouri; Kate McNeive of Scottsdale, Arizona; and Tina Litsey of Kansas City, Missouri.

How Are the First Days of Winter and Summer Chosen?

This Imponderable was posed by a caller on John Dayle's radio show in Cleveland, Ohio. John and the supposed Master of Imponderability looked at each other with blank expressions. Neither one of us had the slightest idea what the answer was. What did it signify?

We received a wonderful answer from Jeff Kanipe, an associate editor at *Astronomy*. His answer is complicated but clear, clearer than we could rephrase. So Jeff generously has consented to let us quote him in full:

> The first day of winter and summer depend on when the sun reaches its greatest angular distance north and south of the celestial equator.

Imagine for a moment that the Earth is reduced to a tiny ball floating in the middle of a transparent sphere and that we're on the "outside" looking in. This sphere, upon which the stars seem fixed and around which the moon, planets, and sun seem to move, is called the celestial sphere. If we simply extend the earth's equator to the celestial sphere it forms a great circle in the sky: the celestial equator.

Now imagine that you're back on the Earth looking out toward the celestial sphere. You can almost visualize the celestial equator against the sky. It forms a great arc that rises above the eastern horizon, extends above the southern horizon, and bends back down to the western horizon.

But the sun doesn't move along the celestial equator. If it did, we'd have one eternal season. Rather, the seasons are caused because the Earth's pole is tilted slightly over 23 degrees from the "straight up" position in the plane of the solar system. Thus, for several months, *one hemisphere tilts toward the sun while the other tilts away*. The sun's apparent annual path in the sky forms yet another great circle in the sky called the ecliptic, which, not surprisingly, is inclined a little over 23 degrees to the celestial equator.

Motions in the solar system run like clockwork. Astronomers can easily predict (to the minute and second!) when the sun will reach its greatest angular distance north of the celestial equator. This day usually occurs about June 21. If you live in the Northern Hemisphere and note the sun's position at noon on this day, you'll see that it's very high in the sky because it's as far north as it will go. The days are longer and the nights are shorter in the Northern Hemisphere. The sun is thus higher in the sky with respect to our horizon, and remains above the horizon for a longer period than it does during the winter months. Conditions are reversed in the Southern Hemisphere: short days, long nights. It's winter there.

Just reverse the conditions on December 22. In the Northern Hemisphere, the sun has moved as far south as it will go. The days are short, while the lucky folks in the Southern Hemisphere are basking in the long, hot, sunny days.

The first days of spring and fall mark the vernal and autumnal equinox, when the sun crosses the equator traveling north and south. As astronomer Alan M. MacRobert points out, the seasonal divisions are rather arbitrary:

> Because climate conditions change continuously, there is no real reason to have four seasons instead of some other number. Some cultures recognize three: winter, growing, and harvest. When I lived in northern Vermont, people spoke of six: winter, mud, spring, summer, fall, and freezeup.

Why Do Most Cars Have Separate Keys for the Ignition and Doors? Doesn't This Policy Increase the Chances of Locking Yourself Out of the Car?

The automakers aren't so concerned about *you* getting into your car. They are worried about thieves getting into your car.

Ford Motor Company, for example, now uses one key for the ignition and doors and a separate key for the glove compartment and trunk. Ford once used the same key for the door and the trunk, but changed. A Ford representative, Paul Preuss, explains:

> At one time, it was a relatively easy matter for a car thief to work open a car door and make an imprint so that it was possible to produce a key that also worked the ignition. Hence, a separate ignition key. Changing from a five-cut key to the present ten-cut key accomplishes the same thing. Five of the cuts activate the door lock and a different five operate the ignition. Taking an imprint of the door lock does not provide the proper cuts for the ignition lock.

General Motors also provides a separate key for doors and ignition and explains its decision as an attempt to foil aspiring thieves.

A two-key approach also allows the car owner to stash valuables in the trunk or glove compartment while leaving only the ignition key with a parking lot attendant or valet. And if you misplace the door key? Well, there's always the coat hanger.

Submitted by Doris Hosack of Garfield Heights, Ohio. Thanks also to Charles F. Myers of Los Altos, California; and Loretta McDonough of Frontenac, Missouri.

P.S. News Flash. Just as this book was going to press, we received a note from the Ford Parts and Service Division. Although the company felt that separate door and ignition keys made sense for security reasons, Ford is returning to its roots: "The consumer prefers one key for both door and ignition; therefore, we will phase in one key for both in the near future."

What's the Difference Between Popcorn and Other Corn? Can Regular Corn Be Popped?

There are five different types of corn: dent, flint, pod, sweet, and popcorn. Popcorn is the only variety that will pop consistently. Gregg Hoffman, of American Popcorn, told *Imponderables* that other corn might pop on occasion but with little regularity.

The key to popcorn's popping ability is, amazingly, water. Each popcorn kernel contains water, which most popcorn processors try to maintain at about a 13.5% level. The water is stored in a small circle of soft starch in each kernel. Surrounding the soft starch is a hard enamel-like starch. When the kernel is heated for popping, the water inside heats and begins to expand. The function of the hard starch is to resist the water as long as possible.

When the water expands with such pressure that the hard starch gives way, the water bursts out, causing the popcorn kernel to explode. The soft starch pops out, and the kernel turns inside out. The water, converted into steam, is released (fogging the eyeglasses of four-eyed popcorn makers), and the corn pops.

DAVID FELDMAN

The other four varieties of corn are able to store water effectively. But their outer starch isn't hard enough to withstand the water pressure of the expanding kernel, and so nothing pops.

Submitted by David Andrews of Dallas, Texas.

What Does It Mean When We Have 20–20 or 20–40 Vision?

The first number in your visual acuity grade is always twenty. That's because the 20 is a reference to the distance you are standing or sitting from the eye chart. The distance is not a coincidence. Rays of light are just about parallel twenty feet from the eye chart, so that the muscle controlling the shape of the lens in a normal eye is in a state of relative rest when viewing the chart. Ideally, your eyes should be operating under optimal conditions during the eye test.

The second number represents the distance at which a normal eye should be able to see the letters on that line. The third from the bottom line on most eye charts is the 20–20 line. If you can see the letters on that line, you have 20–20 ("normal") vision. A higher second number indicates your vision is subnormal. If you have 20–50 vision, you can discern letters that "normal" observers could see from more than twice as far away, fifty feet. If you achieve the highest score on the acuity test, a 20–10, you can spot letters that a normal person could detect only if he were 50% closer.

We also got the answer to another Imponderable we've always had about the vision test: Are you allowed to miss one letter on a line and still get "credit" for it? Yes, all you need to do is identify a majority of the letters on a line to get full credit for reading it. If only our schoolteachers were such easy graders.

WHEN DO FISH SLEEP? 143

HOW Does Yeast Make Bread Rise? Why Do We Need to Knead Most Breads?

Yeast is a small plant in the fungus family (that's ascomycetous fungi of the genus *Saccharomyces*, to you botanical nuts), and as inert as baker's yeast might seem to you in that little packet, it is a living organism. In fact, it works a little like the Blob, feeding and expanding at will.

Yeast manufacturers isolate one healthy, tiny cell, feed it nutrients, and watch it multiply into tons of yeast. One gram of fresh yeast contains about ten billion living yeast cells, thus giving yeast the reputation as the rabbit of the plant world.

To serve the needs of bakers, manufacturers ferment the yeast to produce a more concentrated product. But the yeast isn't satisfied to idly sit by in the fermentation containers—it wants to eat. So yeast is fed its favorite food, molasses, and continues to grow. A representative of Fleischmann's Yeast told *Imponderables* that under ideal conditions, one culture bottle of yeast holding about two hundred grams will grow to about one hundred fifty tons in five days, enough yeast to make about ten million loaves of bread.

After it has grown to bulbous size, the yeast is separated from the molasses and water and centrifuged, washed, and either formed into cakes or dried into the granulated yeast that most consumers buy. When the baker dissolves the yeast in water, it reactivates the fungus and reawakens the yeast's appetite as well.

Yeast loves to eat the sugar and flour in bread dough. As it combines with the sugar, fermentation takes place, converting the sugar into a combination of alcohol and carbon dioxide. The alcohol burns off in the oven, but small bubbles of carbon dioxide form in the bread and are trapped inside the dough. The carbon dioxide gas causes gluten, a natural protein fiber found in flour, to stretch and provide a structure for the rising dough without releasing the gas. When the dough doubles in size, the recommended amount, it is full of gas bubbles and therefore has a lighter consistency than breads baked without yeast.

By kneading the bread, the baker toughens the gluten protein structure in the dough, stretching the gluten sufficiently to withstand the pressure of the expanding carbon dioxide bubbles. You don't need to knead all dough, however; for instance, batter breads, which are made with less flour and have a more open, coarse grain, don't need it.

Submitted by Jim Albert of Cary, North Carolina.

Why Do Doctors Tap on Our Backs During Physical Exams?

We've always been suspicious about this tapping. From a patient's point of view, it has two strong attributes: It doesn't hurt and it doesn't cost anything extra. But nothing ever seems to happen as a result of the tapping. No doctor has ever congratulated us on how great our back sounded or for that matter looked worried after giving us a few whacks on the back. At our most cynical, we've even wondered whether this is a physical examination equivalent of a placebo: The doctor gets a break from the anxious gaze of the patient, and the patient is reassured that at least the back part of his body is O.K.

Doctors insist that there is a sound reason to tap our backs. Short of an X ray, the tap is one of the best ways to collect information about the lungs. The space occupied by the lungs is filled with air. The two lungs are contained in the two pleural spaces, full of air, and lung tissue itself contains air.

Dr. Frank Davidoff, associate executive vice president, Education, for the American College of Physicians, told *Imponderables* about the fascinating history of the practice of tapping:

In 1754, a Viennese physician named Leopold Auenbrugger discovered that if you thumped the patient's chest, it would give off a more hollow sound when you tapped over the air-filled lung space, and a more "flat" or "dull" sound if you tapped over a part of the chest that was filled with something more solid, like muscle, bone, etc. Auenbrugger's father was a tavern keeper in Graz, Austria, who used to judge the amount of wine left in the casks by tapping on them—the hollow note indicating air, the flat note indicating wine.

Auenbrugger found that by thumping a patient's chest—somewhat as his father rapped on a cask—abnormal lesions in the chest cavity, such as fluid or a solid tumor in the cavity where air-filled lung ought to be, produced a sound different from that given off in a healthy air-filled chest. Auenbrugger tested out his new method of physical diagnosis over a period of seven years of drumming on his patient's chests, and in 1761, he put before the medical profession the result of his experiments, in a book called *New Invention to Detect by Percussion Hidden Diseases in the Chest*.

Dr. Davidoff adds that the technique used today is virtually the same as the one Auenbrugger invented more than two hundred years ago.

Dr. William Berman, of the Society for Pediatric Research, says that the technique is a good, obviously cheap alternative to an X ray and has even other attributes. Tapping on the front of the chest can determine the size of a patient's heart, because the heart is much more solid than the lungs as it is muscular and full of blood.

Submitted by Richard Aaron of Toronto, Ontario.

DAVID FELDMAN

Why Do Military Personnel Salute One Another?

Every Western military organization we know of has some form of hand salute. In every culture, it seems the inferior initiates the salute and is obligated to look directly into the face of the superior.

The origins of the hand salute are murky. In ancient Europe, where not only military officers but freemen were allowed to carry arms, the custom for men about to encounter one another was to lift their right hand to indicate they had no intention of using their sword. Many of our friendly gestures, such as tipping hats, waving, and handshaking, probably originated as ways of proving that one's hand was not reaching for a sword or a convenient rock.

By the time of the Roman Empire, salutes were a part of formal procedure among the military. Soldiers saluted by placing their right hands up to about shoulder height with the palm out. The hand never touched the head or headgear during the salute.

In medieval times, when knights wore steel armor that covered their bodies from head to toe, two men often encountered each other on horseback. To display friendship, two knights supposedly would raise their visors, exposing their faces and identities to view. Because they held their reins in the left hand, they saluted with their right (sword) hand, an upward motion not unlike the salute of today.

Whether or not our modern salute stems from the rituals of chivalry, we know for a fact that we Americans borrowed our salute from modern British military practices. In 1796, British Admiral Earl of St. Vincent commanded that all British officers must henceforth take off their hats when receiving an order from a superior "and not to touch them with an air of negligence." Although the British Navy made salutes compulsory, it didn't codify the precise nature of the salute. In many cases, inferiors simply "uncovered" (doffed their caps).

The American military salute has also undergone many changes over the years. At one time, Marines didn't necessarily salute with their right hand, but the hand farthest from the officer being saluted. Even today, there are differences among the branches. Although the Army and Air Force always salute with their right hand, Navy personnel are allowed to salute with the left hand if the right is encumbered. And while Air Force and Army men and women may salute while sitting down, Naval officers are forbidden to do so.

Even if the motivations of ancient saluters were to signal friendly intentions, the gesture over the years has been transformed into a ritual signifying respect, even demanding subjection, and a tool to enforce discipline. The United States Marine Corps, though, has maintained a long tradition of shunning any symbols of servility. In 1804, Marine Commandant William Ward Burrows knowingly discarded the European tradition of inferiors uncovering before superiors and issued this order:

DAVID FELDMAN

No Marine in the future is to take his hat off to any person. When the officer to be saluted approaches, he will halt, face the officer and bring his right hand with a quick motion as high as the hat, the palm in front.

As a Marine publication notes, Burrows' order did much for the esprit de corps:

We can be certain of one fact—the newly initiated salute was popular with enlisted personnel, for an English traveler of that period (Beachey) reported that "the Marines, although civil and well disciplined, boast that they take their hats off to no one."

Submitted by Wally DeVasier of Fairfield, Iowa. Thanks also to George Flower of Alexandria, Virginia.

Why Do Recipes Warn Us Not to Use Fresh Pineapple or Kiwifruit in Gelatin? Why Can We Use Canned Pineapple in Gelatin?

Both pineapple and kiwifruit contain enzymes that literally break down gelatin into a pool of glop. The enzyme in pineapple, papain, is also found in papaya and many other tropical fruits. According to the president of the California Kiwifruit Commission, Mark Houston, kiwifruit contains a related enzyme, actinidin, that similarly breaks down gelatin, preventing jelling.

Papain is a particularly important enzyme that has more functions than turning your Jell-O mold into a Jell-O pool. Papain is the active ingredient in meat tenderizers. Just as papain splits the protein in gelatin, it also attacks proteins in meat. Ever experience a stinging sensation in your mouth while eating a fresh pineapple? Papain is attacking your throat.

How can we contain this rapacious enzyme? Just as Kryptonite incapacitates Superman or garlic renders Dracula useless, so heat is the enemy of protein-splitting enzymes such as papain or actinidin. Canned pineapple can be used effectively in gelatin because the heat necessary to the process of canning fruit inactivates the enzymes. Canned pineapple might not taste as good as fresh, but it is much easier on the throat.

Submitted by Marsha Beilsmith of St. Charles, Missouri. Thanks also to David Freling of Hayward, California; and Susan Stock of Marlboro, Massachusetts.

Where Is Donald Duck's Brother?

"We see Donald Duck's nephews, Huey, Dewey, and Louie, but we never see their Dad, Donald's brother. Why not?" wails our concerned correspondent.

The main reason we never see Donald's brother is that he doesn't have one. He does have a sister with the infelicitous name of Dumbella. In a 1938 animated short, *Donald's Nephews*, Donald receives a postcard from his sister informing him that she is sending her "three angel children" for a visit.

Poor Donald, excitedly anticipating the arrival of Masters Huey, Dewey, and Louie, had no idea either that the little visit would turn into a permanent arrangement or, since his sister really thought they were little angels, that she had really earned her name. The three ducklings, indistinguishable in their personalities and equally adept in their propensity for mischief, continued to torture Donald and Scrooge McDuck in many cartoon shorts.

In a 1942 short, *The New Spirit*, Donald lists the three dependents in a tax form as adopted, indicating that Donald was a most generous brother, a certified masochist, and just as dumb as Dumbella.

Submitted by Karen S. Harris of Seattle, Washington.

DAVID FELDMAN

What Causes Bags Under the Eyes?

Let us count the ways, in descending order of frequency:

1. Heredity. That's right. It wasn't that night on the town that makes you look like a raccoon in the morning. It's all your parents and grandparents' fault. Some people are born with excess fatty tissue and liquid around the eyes.
2. Fluid retention. The eyelids are the thinnest and softest skin in the entire body, four times as thin as "average" skin. Fluid tends to pool in thin portions of the skin.

    What causes the fluid retention? Among the culprits are drugs, kidney or liver problems, salt intake, and very commonly, allergies. Cosmetics drum up more business for dermatologists and allergists than just about anything else. Allergic reactions to mascara and eyeliner are the usual culprits.
3. Aging. The skin of the face, particularly around the eyes, loosens with age. Age is more likely to cause bags than mere sleepiness or fatigue.
4. Too many smiles and frowns. These expressions not only can build crow's feet but bags. We can safely disregard this answer to explain Bob Newhart's bags, however.

Another less fascinating explanation for many sightings of bags under the eyes was noted by Dr. Tom Meek, of the American Academy of Dermatology, in the *New York Times:* "The circles are probably caused by shadows cast from overhead lighting. . . ."

Submitted by Stephen T. Kelly of New York, New York.

How Do Blind People Discriminate Between Different Denominations of Paper Money?

Sandra Abrams, supervisor of Independent Living Services for Associated Services of the Blind, points out that the government defines "legally blind" as possessing 10% or less of normal vision. Legally blind people with partial vision usually have few problems handling paper money:

> Individuals who are partially sighted may be able to see the numbers on bills, especially in certain lighting conditions. Some people with low vision must hold the money up to their noses in order to see the numbers; some people have been asked by members of the public if they are smelling their money. Other persons with low vision might use different types of magnification. Some people with partial sight have pointed out that the numbers on the top corners of bills are larger than those on the bottoms.

The U.S. government certainly doesn't make it easy for blind people to identify currency. Virtually every other nation varies the size and color of denominations. One reader asked

DAVID FELDMAN

whether a five-dollar bill *feels* different from a twenty-dollar bill. Although suggestions have been made to introduce slight differences in texture, a blind person can't now discriminate between bills by touching them.

Initially blind people must rely on bank tellers or friends to identify the denomination of each bill, and then they develop a system to keep track of which bill is which. Gwynn Luxton, of the American Foundation for the Blind, uses a popular system with her clients:

- One-dollar bills are kept flat in the wallet.
- Five-dollar bills are folded in half crosswise, so that they are appproximately three inches long.
- Ten-dollar bills are folded in thirds crosswise, so that they are approximately two inches long.
- Twenty-dollar bills are folded in half lengthwise, so that they are half the height of the other bills and sit down much farther in the wallet or purse than the other bills.

Machines have been created to solve this problem as well. The relatively inexpensive Talking Wallet reads out the denomination of bills it receives. The more expensive Talking Money Identifier can be hooked up to cash registers and be used for commercial use. Many newspaper vendors are blind, and the Money Identifier can save them from being shortchanged.

Blind people have so many pressing problems imposed on them by a seeing culture that identifying paper money is a minor irritant. As Sandra Abrams puts it, "Frankly, of all the things I do daily, identifying money is one of the easiest."

Submitted by Jon Gregerson of Marshall, Michigan.

When Not Flying, Why Do Some Birds Walk and Others Hop?

Birds are one of the few vertebrates that are built for both walking and flying. Physiologically, flying is much more taxing on the body than walking. Usually a bird without fear of attack by predators in its native habitat will eventually become flightless. New Zealand, an oceanic island with few predators, has flightless cormorants, grebes, wrens, and even a flightless owl parrot. As Joel Carl Welty states in *The Life of Birds:*

> Why maintain splendid wings if the legs can do an adequate job? This principle may well explain why birds who are good runners fly poorly or not at all. And some of the best fliers, such as hummingbirds, swifts, and swallows, are all but helpless on their feet.
>
> More birds are hoppers than walkers. Birds that walk or run characteristically possess long legs and live in wide open spaces. While the typical tree dweller has four toes on each foot, many walkers have only two or three. Most tree-dwelling birds are hoppers, because it is easier to navigate from branch to branch by hopping than by walking. Most birds that hop in trees will hop on the ground. Although each hop covers more ground than a step would, the hop is more physically taxing.

Dr. Robert Altman, of the A & A Veterinary Hospital, points out that some birds will hop or walk depending on the amount of ground they plan to cover. "For a few steps, it might be easier for a bird to hop from place to place as he would from perch to perch in trees. To cover longer distances, the bird would walk or run."

Submitted by Jill Clark of West Lafayette, Indiana.

DAVID FELDMAN

Why Does String Cheese "String" When Torn Apart?

If you read *Why Do Clocks Run Clockwise? and Other Imponderables,* and shame on you if you haven't, you know that newspapers tear easily in a vertical position because all the fibers are lined up in the same direction when pulp is put into the paper-making machine. String cheese works on exactly the same principle.

When producing string cheese, the cheese curd is formed into a large mass and then stretched mechanically. The stretching causes the protein fibers to line up in a parallel fashion. According to Tamara J. Hartweg, of Kraft, "This physical modification of the protein structure is what causes the stringing quality of the cheese. When peeled, the protein fibers, which are aligned in one direction, come off in strings."

Submitted by Lee Hand of Newbury Park, California.

WHEN DO FISH SLEEP?

Who Got the Idea of Making Horseshoes? Why are Horse-shoes Necessary? What Would Happen If Horses Weren't Shod?

If horses weren't shod, they would probably have trouble getting served at fast-food establishments. Maybe they can get away with no shirts. But no shoes?

But seriously, folks, horses have the Romans to blame for the end of their barefoot existence. Horses were perfectly happy galloping around without shoes until the leaders of the Roman Empire decided that it would be a good idea to build paved roads. Without support, horses' hooves would split and crack on the hard pavement.

The paving of roadways hastened the time when horses, used to riding the range in the wild, were domesticated and forced to carry loads and pull heavy carts. These added burdens put strain on horses' feet, so the Romans used straw pads as the first horseshoes.

Karen L. Glaske, executive secretary of the United Professional Horsemen's Association, says that although evolution has bred out some of the toughness of horses' feet, many can still live a barefoot life:

> Shoes are not essential to a horse that is left to pasture or used only as an occasional trail mount. However, the stresses which horses' feet endure when jumping, racing, showing, or driving make it necessary for the conscientious owner to shoe the animal. It is a protective measure.

DAVID FELDMAN

Why Are Tattoos Usually Blue (With an Occasional Touch of Red)?

Most tattoos are not blue. The pigment, made from carbon, is actually jet black. Since the pigment is lodged *underneath* the skin, tattoos appear blue because of the juxtaposition of black against the yellowish to brown skin of most Caucasians. Although red is the second most popular color, many other shades are readily available; in fact, most tattoo artists buy many different colorings, premade, from Du Pont.

We spoke to Spider Webb, perhaps the most famous tattooist in the United States and leader of the Tattoo Club of America, about the prevalence of black pigment in tattoos. Webb felt that most clients, once they decide to take the plunge, want to show off their tattoos: Black is by far the strongest and most visible color. Webb added that in the case of one client, albino guitarist Johnny Winter, a black tattoo does appear to be black and not blue.

Submitted by Venia Stanley of Albuquerque, New Mexico

Why Is the Width of Standard Gauge Railroads Four Feet Eight-and-One-Half Inches?

When tramways were built in England to carry coal by cart or coach, the vehicles were built with wheels four feet eight-and-one-half inches apart. Legend has it that this was the same distance apart as Roman chariot wheels, but we doubt it for one important reason: There is a more logical explanation. Track gauges are determined by measuring from the *inside* of one rail to the *inside* of the other. However, the rails themselves occupied three-and-one-half inches of space. In other words, fifty-six-

and-one-half inches was almost certainly derived by starting with a measurement of five feet and deducting the width of the rails themselves.

When steam railroads were later constructed in England, the tramway gauge was retained for the most part, and in 1840 Parliament made it official, decreeing four feet eight-and-one-half inches as the standard gauge in Great Britain.

If only the United States were as logical. The first railroad in America, in Massachusetts, featured locomotives from England, built for standard gauge tracks, so the U.S. started with the same track dimensions. But no one in the fledgling American rail industry seemed to consider that it might be nice to have an interlocking system of compatible railways.

As companies from different states started their own lines, anarchy ruled. The Mohawk & Hudson stretched the standard gauge only one half inch, but the Delaware & Hudson featured a six-foot behemoth. In the early and mid-nineteenth century, gauges ranged between a little more than three feet to more than six feet.

Faced with incompatible rolling stock, long delays were common, yet to be preferred to the numerous accidents that ensued when engineers tried to roll locomotives on gauges a few inches too wide at the usual breakneck speeds.

When Union Pacific was about to be built, Abraham Lincoln tried to fix five feet—then the most popular width in the South and California—as the standard gauge for the whole country. But the established railroads in the North and the East objected on financial grounds and managed to lobby to retain fifty-six-and-one-half inches as the standard.

According to railroad expert Alvin Harlow in "The Tangle of Gauges,"

> In 1871 there were no fewer than twenty-three gages, ranging from 3 feet up to 6 feet on the railroads of the United States. Less than fifteen years later there were twenty-five; a considerable group of roads in Maine had been born only two feet wide, whilst

DAVID FELDMAN

a logging company in Oregon had built one that sprawled over 8 feet of ground.

The proliferation of gauges was caused not only by regional stubbornness but because no railroad company seemed willing to spring for the cost of converting its tracks. Finally, Illinois Central broke the logjam. In one wild, torchlit night, Illinois Central workers narrowed six hundred miles of track. Southern railroad companies, reluctant to adopt the Yankee standard, followed suit years later.

Even more difficult than relaying track was the task of refitting the rolling stock. Locomotives and cars were dragged into shops all along their routes. Harlow mentions that although the companies tried to return cars to their home lines for conversion, the logistics were a nightmare. Usually cars were converted wherever they were when the tracks were remodeled. Sufficient numbers of new workers had to be hired temporarily to have crews working twenty-four hours a day resetting locomotive truck wheels, removing the tires from truck wheels, and resetting them for the standard gauge.

A few gauges with oddball widths survived into the twentieth century, mostly in New England and the Pacific Northwest, but they were anomalies. The United States eventually rejected the "new and improved" and returned to the standard gauge of the English.

Why Is the Bathtub Drain Right Below the Faucet? Why Isn't the Bathtub Drain on the Opposite Side of the Bathtub from the Faucet?

"Wouldn't this configuration be easier for rinsing purposes?" asks our correspondent Pam Lebo. No doubt it would, but there are plenty of reasons why the plumbing industry is going to continue to make you and the makers of Woolite unhappy.

Now hard as it may be to believe, some people actually use the bathtub for bathing. These heathens would not appreciate having to sit on the drain (or for that matter, having the spigot clawing at their backs). John Laughton, of American Standard, raises another legitimate objection: A dripping faucet in Pam's configuration would cause a stain on the whole length of the bathtub.

Your dream configuration would have other practical drawbacks. Peter J. Fetterer, of Kohler Company, explains why:

> The bathtub drain is generally at the same location as the water supply because of the piping required for both. Drains and supplies run through buildings in plumbing chases, vertical spaces for pipes that move water from floor to floor. Drains are attached to vent pipes that run through the chases and vent to the outside of a structure. These chases use up living space and are kept to a minimum for economic reasons.

So must we resign ourselves to a lifetime of boring bathtubs? Not necessarily. Pam's configuration might attract some who take only showers, but it will probably never be popular. However, American Standard has created a bathtub that presents interesting possibilities for extracurricular activities besides rinsing. Their avant garde bathtub places both the faucet and the drain halfway along the bath with, offers John Laughton, "a back slope at both ends so that two could bathe together in comfort and save water." Save water. Sure, Mr. Laughton.

Submitted by Pam Lebo of Glen Burnie, Maryland.

Do Fish Sleep? If So, When Do Fish Sleep?

Our trusty *Webster's New World Dictionary* defines sleep as "a natural, regularly recurring condition of rest for the body and mind, during which the *eyes are usually closed* and there is *little or no conscious thought or voluntary movement.*" Those strategically placed little weasel words we have italicized make it hard for us to give you a yes or no answer to this mystery. So as much as we want to present you with a tidy solution to our title Imponderable, we feel you deserve the hard truth.

Webster probably didn't have fish in mind when he wrote this definition of "sleep." First of all, except for elasmobranchs (fish with cartilaginous skeletons, such as sharks and rays), fish don't have eyelids. So they can't very well close them to sleep. No fish has opaque eyelids that block out vision, but some have a transparent membrane that protects their eyes from irritants.

Pelagic fish (who live in the open sea, as opposed to coasts), such as tuna, bluefish, and marlins, *never* stop swimming. Jane Fonda would be proud. Even coastal fish, who catch a wink or two, do not fall asleep in the same way humans do. Gerry Carr, director of Species Research for the International Game Fish Association, wrote us about some of the ingenious ways that fish try to catch a few winks, even if forty winks are an elusive dream:

Some reef fishes simply become inactive and hover around like they're sleeping, but they are still acutely aware of danger approaching. Others, like some parrot fishes and wrasses, exude a mucus membrane at night that completely covers their body as though they've been placed in baggies. They wedge themselves into a crevice in the reef, bag themselves, and remain there, semi-comatose, through the night. Their eyes remain open, but a scuba diver can approach them and, if careful, even pick them up at night, as I have done. A sudden flurry of movement, though, will send them scurrying. They are not totally unaware of danger.

In many ways, fish sleep the same way we plod through our everyday lives when we are awake. Our eyes are open but we choose, unconsciously, not to register in our brains most of the sensory data we see. A fish sleeping is in a state similar to the poor fish depicted watching the slide show in Kassie Schwan's illustration. We stare at the screen with our eyes open, but our minds turn to mush. If a crazed assassin burst into the room, we could rouse ourselves to attention, but if someone asked us to describe what fabulous tourist attraction we were watching, we couldn't say whether it was Stonehenge or the Blarney Stone.

If you accept that a fish's blanking out is sleeping, then the answer to the second part of the mystery is that fish sleep at night, presumably because of the darkness. Anyone with an aquarium can see that fish can float effortlessly while sleeping. They exude grace—which is more than we can say for how most humans look when they are sleeping.

Submitted by Karole Rathouz of Mehlville, Missouri. Thanks also to Cindy and Sandor Keri of Woodstock, Georgia; and Heather Bowser of Tulsa, Oklahoma.

DAVID FELDMAN

Why Do We Seem to Feel Worse at Night When We Have a Cold?

For the same reason that your feet swell up and hurt after a long day standing up. To quote Dr. Ernst Zander, of Winthrop Consumer Products:

> Nasal obstruction, produced by a great variety of conditions, usually seems worse to a patient when he is lying down. This is because tissue fluids and blood tend to pool in the head more when he is recumbent than when he is standing.

Of course, one is generally more likely to feel tired and worn out at night. But the doctors who *Imponderables* consulted indicated that reclining for long periods of time will worsen symptoms—one reason why often we feel lousy despite the "luxury" of being able to lie in bed all day long when we are sick.

Why Do Many Dry Cleaning Stores Advertise Themselves as "French" Dry Cleaners? Is There Any Difference Between a French Dry Cleaner and a Regular Dry Cleaner?

To answer the last part of this Imponderable first, there is a BIG difference between a French dry cleaner and a regular dry cleaner: about one dollar per garment.

Sure, some justification exists for calling any dry cleaning establishment "French." Dry cleaning was supposedly discovered in the 1830s by one Jolie Belin, a Frenchman who reputedly tipped over a kerosene lamp on a soiled tablecloth and found that the oil eliminated the stains. The story of Jolie Belin might be apocryphal, but dry cleaning definitely started in France.

Most Yankees are so cowed by the image of anyone who can speak French and order fancy wines in restaurants that we not only entrust our best clothing to them but are willing to pay extra for the artistry of the French dry cleaner.

DAVID FELDMAN

We conveniently forget, though, that the owner of the French dry cleaning store is as likely to be Japanese as French. And the French dry cleaner is unlikely to tell you that there is absolutely no difference between the way he and the One Hour Martinizing store down the block cleans your clothes.

Submitted by Mrs. Shirley Keller of Great Neck, New York.

Why Do Kellogg's Rice Krispies "Snap! Crackle! and Pop!"?

Kellogg's Rice Krispies have snapped, crackled, and popped since 1928. Kellogg's production and cooking process explains the unique sound effects.

Milled rice, from which the bran and germ have been removed, is combined with malt flavoring, salt, sugar, vitamins, and minerals and then steamed in a rotating cooker. The rice, now cooked, is left to dry and temper (i.e., sit while the moisture equalizes). The rice is then flattened and flaked as it passes through two cylindrical steel rollers. The Krispies are left to dry and temper for several more hours.

The cereal then moves to a toasting oven. The flattened rice is now exposed to hot air that puffs each kernel to several times its original size and toasts it to a crisp consistency. This hot air produces tiny air bubbles in each puff, crucial in creating the texture of Rice Krispies and their unique sound in the bowl.

When milk is added to the prepared cereal, the liquid is unevenly absorbed by the puffs, causing a swelling of the starch structure. According to Kellogg's, "This swelling places a strain on the remaining crisp portion, breaking down some of the starch structure and producing the famous 'Snap! Crackle! and Pop!' "

Submitted by Kevin Madden of Annandale, New Jersey.

WHEN DO FISH SLEEP?

Why Do So Many Cough Medicines Contain Alcohol?

No, the alcohol isn't there to make you forget the taste of the cough medicine. *Nothing* could do that.

Some drugs don't mix well with water. Alcohol is the best substitute. Although the alcohol may help some people sleep, the alcohol in the recommended doses of most cough medicines isn't high enough to affect the average person (one teaspoon has less than 10% the alcohol of a shot of whiskey).

Why Do Letters Sent First Class Usually Arrive at Their Destination Sooner than Packages Sent by Priority Mail?

When we send a package through the United States Postal System, we have alternatives. We can send them third class (and for certain goods, fourth class) for considerably less than Priority Mail, the package equivalent of first-class mail. But our experience is that packages invariably take longer to arrive. So we asked the USPS why. Their answers:

1. Packages are canceled and processed by hand. Almost all letters are canceled and processed by machines. Letters are sorted by OCR (Optical Character Reader) machines capable of processing up to thirty thousand letters in one hour. These machines "read" the last line of the address and sort the envelopes by zip code. Even if the OCRs can't read a letter, another machine helps humans to do so. The letter is transferred to an LSM (Letter Sorting Machine), which pops up a letter one second at a time before a postal worker who routes the letter to the proper zip code.
2. Samuel Klein, public affairs officer of the United States Postal Service, says that if a package is larger than a shoe box or weighs more than two pounds, it must be delivered by a parcel-post truck, which also carries nonpriority packages.

DAVID FELDMAN

3. Postal workers inadvertently treat Priority Mail as fourth-class mail. Dianne V. Patterson, of the Office of Consumer Affairs of the USPS, warns that "If the Priority Mail or First-Class stamps or stickers are not prominently placed on the parcel, it stands a good chance of being treated as fourth-class mail."

It isn't hard to understand the tremendous logistical difficulties in delivering mail across a large country, or even why mail might be delivered more slowly than we would like. But it is hard to understand exactly how the post office discriminates between processing a first-class and a fourth-class delivery. In the days when airmail was a premium service and fourth-class mail was transported by rail, we understood the distinction. But are postal workers now encouraged to malinger when processing fourth-class mail? Are they taught to let it sit around delivery stations for a few days so as not to encourage customers to use the slower service?

Despite our grumbling, we've found the USPS to be dependable in delivering all the free books we sent out to Imponderables posers. But we'll share a nasty secret. The books we send out at Special Fourth Class (book rate) seem to arrive no later than the books we send by the costlier Priority Mail.

Why Isn't There a Holiday to Commemorate the End of the Civil War?

Reader Daniel Marcus, who sent in this Imponderable, stated the mystery well:

> We observe a national holiday to commemorate the end of World War I on November 11 [Veteran's Day], and newspapers always note the anniversaries of V-E and V-J Days regarding the end of World War II. The Revolutionary War is honored, of course, on July 4. Why isn't there a national holiday to celebrate the end of the Civil War, the second most important and only all-American war in our history?

Good question, Daniel, but one that assumes a false premise. Memorial Day (also known as Decoration Day), celebrated on the last Monday of May, now honors the dead servicemen and servicewomen of all wars. But originally it honored the Civil War dead.

In his book *Celebrations*, historian Robert J. Myers credits Henry C. Welles, a druggist in Waterloo, New York, for originat-

DAVID FELDMAN

ing the idea of decorating the graves of dead Civil War veterans in 1866. Originally the holiday was celebrated on May 5, when townspeople would lay flowers on the servicemen's graves.

John A. Logan, commander in chief of the Grand Army of the Republic (a veterans' support group), declared in 1868 that Decoration Day should be observed throughout the country. New York State was the first to make the day a legal holiday in 1873. Although Memorial Day never officially became a national holiday, it is celebrated in almost every state on the last Monday in May.

As with most holidays, the average person does not necessarily celebrate the occasion with the solemnity the founders of the holiday envisioned. In his study of the Civil War era, *The Expansion of Everyday Life, 1860–1878*, historian Daniel E. Sutherland notes that the new Memorial Day conveniently filled the void left by the declining popularity of George Washington's birthday: "Brass bands, picnic lunches, baseball games, and general merrymaking soon attached themselves to the new holiday, as it became as much a celebration of spring as a commemoration of the nation's honored dead." Today, the holiday is more often viewed as a kickoff to summertime than a serious tribute to the war dead.

Southerners, as might be expected, didn't particularly cotton to the concept of the northerner's Memorial Day. They countered with Confederate memorial days to honor their casualties, and many southern states still observe these holidays today. Florida and Georgia's Confederate Memorial Day is April 26; and Alabama and Mississippi celebrate on the last Monday of April. Not coincidentally, the president of the Confederacy, Jefferson Davis, was born on June 3. Kentucky and Louisiana celebrate the day as a state holiday.

Submitted by Daniel Marcus of Watertown, Massachusetts.

Is It True that Permanents Don't Work Effectively on Pregnant Women?

No, it isn't true, despite the fact that our correspondent has been told that it *is* true by her hairdressers. And you are not alone; we have been asked this Imponderable many times.

Everett G. McDonough, Ph.D., senior vice president of Zotos International, Inc., is one of the pioneers of permanent waving (he has worked at Zotos since 1927), and he is emphatic. He has seen or read the results of fifty thousand to one-hundred thousand perms given in the Zotos laboratory over the past sixty years. He has never seen the slightest evidence that pregnancy has any effect on permanent waving. And for good reason:

> a hair fibre after it emerges from the skin has no biological activity. Whether it remains attached to the scalp or is cut off, its chemical composition will remain the same. In either case the chemical composition can be altered only by some external means.

Louise Cotter, consultant to the National Cosmetology Association, reiterated McDonough's position and explained how a permanent wave actually works.

> A hair is held together by a protein helix consisting of salt, hydrogen, and disulphide bonds. The words "permanent wave" refer to the chemical change that takes place when those bonds are broken by a reducing agent having a pH of 9.2. The hair, when sufficiently softened, is re-bonded (neutralized) with a solution having a pH of 7.0–7.9. This causes the hair to take the shape of the circular rod on which it is wound, creating full circle curls or a wave pattern, depending upon the size and shape of the rod.

Although Cotter says that poor blood circulation, emotional disturbances, malfunctioning endocrine glands, and certain drugs may adversely affect the health of hair, none of these factors should alter the effectiveness of a perm on a pregnant woman. Pregnancy isn't an illness, and none of these four factors is more likely in pregnant women. Even if a pregnant woman takes hormones that could conceivably affect the results of a perm, a cosmetologist can easily compensate for the problem.

John Jay, president of Intercoiffure, answers this Imponderable simply:

> I have never had a permanent-wave failure due to pregnancy. Should failure occur for whatever reason, pregnancy may be the most convenient excuse available to some hairdressers.

Submitted by Jeri Bitney of Shell Lake, Wisconsin.

Why Do Some Escalator Rails Run at a Different Speed from the Steps Alongside Them?

The drive wheel that powers the steps in an escalator is attached to a wheel that runs the handrails. Because the steps and the rails run in a continuous loop, the descending halves of the stairs and handrails act as a counterweight to their respective ascending halves. The handrails, then, are totally friction-driven rather than motor-driven.

If the escalator is properly maintained, the handrail should move at the same speed as the steps. The handrails are meant to provide a stabilizing force for the passenger and are thus designed to move synchronously for safety reasons. Handrails that move slower than the accompanying steps are actually dangerous, for they give a passenger the impression that his feet are being swept in front of him. Richard Heistchel, of Schinder Elevator Company, informed *Imponderables* that handrails were once set to move slightly faster than the steps, because it was believed that passengers forced to lean forward were less likely to fall down.

> *Submitted by John Garry, WTAE Radio, Pittsburgh,*
> *Pennsylvania. Thanks also to Jon Blees of Sacramento,*
> *California; Robert A. Ciero, Sr. of Bloomsburg, Pennsylvania;*
> *and David Fuller of East Hartford, Connecticut.*

Why Are There Lights Underneath the Bottom Steps of Escalators and Why Are They Green?

Those emerald lights are there to outline the periphery of the step on which you are about to hop or hop off. The majority of accidents on escalators occur when a passenger missteps upon entering or exiting the escalator. The lights, which are located just below the first step of ascending stairs (and the last step of descending stairs), are there to show the way for the unproficient escalator passenger.

Escalator lights are green for the same reason that traffic lights use green: Green is among the most visually arresting colors.

Submitted by John T. Hunt of Pittsburgh, Pennsylvania.

Why Are Rented Bowling Shoes So Ugly?

We know that taste in art is a subjective matter. We are aware that whole books have been written about what colors best reflect our personalities and which colors go best with particular skin tones.

But on some things a civilized society must agree. And rented bowling shoes *are* ugly. Does anybody actually believe that maroon-blue-and-tan shoes best complement the light wood grain of bowling lanes or the black rubber of bowling balls?

Bruce Pluckhahn, curator of the National Bowling Hall of Fame and Museum, told us that at one time "the black shoe—like the black ball—was all that any self-respecting bowler would be caught dead using." Now, most rented bowling shoes are tricolored. The poor kegler is more likely to be dressed like Cindy Lauper (on a bad day) than Don Carter.

We spoke to several shoe manufacturers who all agreed that their three-tone shoes were not meant to be aesthetic delights. The weird color combinations are designed to discourage theft. First, the colors are so garish, so ugly, that nobody *wants* to steal

WHEN DO FISH SLEEP?

them. And second, if the rare pervert does try to abscond with the shoes, the colors are so blaring and recognizable that there is a good chance to foil the thief.

Of course, rented bowling shoes get abused daily. A bowling proprietor is lucky if a pair lasts a year. Gordon W. Murrey, president of bowling supply company Murrey International, told *Imponderables* that the average rental shoe costs a bowling center proprietor about $10 to $25 a pair. The best shoes may get rented five hundred times before falling apart, at a very profitable $1 per rental.

Even if rentals were a dignified shade of brown, instead of black, tan, and red, they would get scuffed and bruised just the same. Bowlers don't expect fine Corinthian leather. But can't the rented bowling shoes look a littler classier, guys? Isn't a huge 9 on the back of the heel enough to discourage most folks from stealing a shoe?

Submitted by Shane Coswith of Reno, Nevada.

What's the Difference Between Virgin Olive Oil and *Extra* Virgin Olive Oil?

We promised ourselves that we wouldn't make any jokes about virgins being hard to find and extra-virgins being impossible to find, so we won't. We will keep a totally straight face while answering this important culinary Imponderable.

We may have trouble negotiating arms reductions, but on one issue the nations of the world agree; thus, the International Olive and Olive Oil Agreement of 1986. This agreement defines the terms "virgin olive oil" and "extra virgin olive oil."

Any olive oil that wants to call itself virgin must be obtained from the fruit of the olive tree solely by mechanical or other physical means rather than by a heating process. The oil cannot be refined or diluted, but may be washed, decanted, and filtered.

DAVID FELDMAN

The lowest grade of virgin olive oil is semi-fine virgin olive oil, which is sold in stores as "virgin." This oil must be judged to have a good flavor and no more than three grams of free oleic acid per hundred grams of oil.

The next highest grade, fine virgin olive oil, cannot exceed one and a half grams of oleic acid per hundred grams and must have excellent taste.

Extra virgin olive oil must have "absolutely perfect flavour" and maximum acidity of one gram per hundred grams. According to José Luis Perez Sanchez, commercial counselor of the Embassy of Spain, extra virgin olives are often used with different kinds of natural flavors and are quite expensive, which any trip to the local gourmet emporium will affirm.

As with many other food items, the prize commodity (extra virgin olive oil) is the one that achieves quality by omission. By being free of extraneous flavors or high acidity, the "special" olive oil is the one that manages what wouldn't seem like too difficult a task: to taste like olives.

Submitted by Phyllis M. Dunlap of St. George, Utah.

Why Are There Cracks on Sidewalks at Regular Intervals? What Causes the Irregular Cracking on Sidewalks?

Believe it or not, those regularly spaced cracks are there to prevent the formation of irregular cracks.

We tend to see concrete as lifeless and inert, but it is not. Concrete is highly sensitive to changes in temperature. When a sidewalk is exposed to a cool temperature, it wants to contract.

Gerald F. Voigt, director of Engineering–Education and Research at the American Concrete Pavement Association, explains that concrete is very strong in compression but only one tenth as strong in tension.

> It would be much easier to break a piece of concrete by pulling on two opposite ends, rather than push it together. Cracking in concrete is almost always caused by some form of tensile development.
>
> In many cases the concrete slabs are restricted by the friction of the base on which they were constructed. This frictional resis-

DAVID FELDMAN

tance will put the slabs in tension as they contract; if the resistance is greater than the tensile strength of the concrete, a crack will form. Something has to give.

Without any form of restraint, the concrete will not crack.

Concrete tends to shrink as it dries, and tends to gain strength over time. Thus, sidewalks are most vulnerable to cracking the first night after the concrete is placed. Two strategies are employed to combat cracking.

Arthur J. Mullkoff, staff engineer at the American Concrete Institute, told *Imponderables* that properly positioned reinforcing steel is often used to reduce cracks. But the most effective method of minimizing cracks is to predetermine where the cracks will be located by installing joints in between segments of concrete.

Those spaces that threaten the well-being of your mother's back are a form of "tooled joints," strategically positioned cracks. These joints are placed in all types of concrete slabs. Gerald Voigt elaborates:

> The concrete is sawed or tooled to approximately one-quarter of the thickness of the slab, which creates a "weakened plane." The concrete will crack through the "weakened plane" joint, because that joint is not as strong. As you can see on almost any sidewalk, tooled joints are placed about every four to eight feet. These joints are placed to control where cracks develop and avoid random cracking which is usually considered unattractive. . . . Typically sidewalks are four inches thick; joint depth must be at least one-quarter of the sidewalk thickness.

Perhaps the most surprising element in the story of concrete cracks is that although so much effort is put into preventing them, cracks are not particularly troublesome. The National Ready-Mixed Concrete Association says that cracks rarely affect the structural integrity of concrete. Even when the cracks are wide enough to allow water to seep in, "they do not lead to progressive deterioration. They are simply unsightly."

Incidentally, our correspondent asked how the superstition "Step on a crack, break your mother's back" originated. We've never found a convincing answer to this Imponderable, but Gerald Voigt offered a fascinating theory:

> Since a concrete sidewalk consists of many short segments (slabs) of white concrete, it can be imagined that it is like a human spine. The spine also consists of many short segments (vertebrae) of white bones. The weak links in each system are the joints. Stepping on a sidewalk crack, or joint, is analogous to stepping on the weakest area of the spine. I imagine if I were walking down a spine, I would avoid stepping on a vertebrae link, wouldn't you?

Submitted by Mrs. Harold Feinstein of Skokie, Illinois.
Thanks also to Henry J. Stark of Montgomery, New York.

Why Do We Have to Shake Deodorant and Other Aerosol Cans Before Using?

If you could see inside a can of deodorant, you would see that the ingredients are not arranged uniformly in the can. The propellant is not soluble and so won't mix with the active ingredients in the deodorant.

In many cases, you would see three or four levels of ingredients in a can. The top layer would contain the hydrocarbon gas used as a propellant. Other active ingredients, such as aluminum salt, emollient, and fragrance, also might seek their own level. By shaking up the can, you would guarantee spraying the proper proportion of ingredients.

Any effort expended in shaking the can is well worth the appreciation from friends and loved ones. But a stiff spray of hydrocarbon gas simply isn't sufficient to take care of a nasty body odor problem.

Submitted by Mark Fusco of Northford, Connecticut.

DAVID FELDMAN

Why Do Airlines Use Red Carbons on Their Tickets?

The dominant manufacturer of airline tickets is Rand McNally, the same company that makes maps and atlases. We spoke to Chris George, of Rand McNally's Ticket Division, who told us that there are two explanations for the tradition of red carbons.

In the early years of commercial aviation, black carbons were used. This we know for a fact. But Mr. George says the problem with black carbons was that in high humidity specks of black would fall off the ticket. Women, in particular, were upset that their hands or gloves were befouled with black crud. So the airlines did market research that revealed women did not object as much to traces of red on their hands because they were used to rouge and lipstick stains. This, Mr. George adds wryly, is the romantic explanation.

The unromantic explanation (a.k.a. the truth) is as follows: Once your ticket form is torn by the ticket agent, it is sent to the accounting department of the airline. The major carriers have long used optical scanners to read the serial numbers found on each ticket. An OCR (Optical Character Recognition) scanner can't read the ticket when black flecks of carbon land on the serial number because it can only register information printed in black ink. Much as a photocopier will not read blue ink, an OCR scanner won't read red ink. Who would have thought that accountants would be responsible for the daring flash of red on airline tickets?

In a time of high-tech stationery, why don't the airlines use carbonless paper? Part of the answer again relates to the OCR equipment. Carbonless paper contains blue specks that OCRs won't read. Furthermore, with chemically sensitized noncarbon paper, legibility is good for only about five copies. Old-fashioned carbon paper can render nine legible copies, sometimes necessary for the daunting itineraries of business travelers.

Now that most airlines are issuing automated ticket boarding passes—the ones that look like computer cards—the decline of the carbonized form is inevitable. Because not all ticket counters possess the equipment to issue these boarding passes, Mr. George predicts that the beloved red carbonized forms will continue to play a part in aviation for the foreseeable future.

Submitted by Niel Lynch of Escondido, California.

I Have a Dollar Bill with an Asterisk After the Serial Number: Is It Counterfeit?

The *Imponderables* staff will gladly accept your dollar bill if you don't want it. No, it's not counterfeit. You are holding a "star note," a replacement for a defective bill that has been destroyed.

In 1910, the Bureau of Engraving and Printing started printing ★ B and later ★ D as prefixes before the serial numbers of replacement notes. No star notes were issued for national bank notes, which were replaced by new notes that matched the missing serial numbers.

Now that notes are issued in series of one hundred million at a time, it is obvious why the Bureau would rather not have to renumber replacement notes, especially since, as Bob Cochran, secretary of the Society of Paper Money Collectors, told us, errors are quite common in the printing process:

> The most common errors are in inking, cutting, and in the overprinting operation. With inking there can be too much, not enough, or unacceptable smears. Notes are printed in sheets of 32; the back is printed in all green ink and then the face is printed in all black ink. If one side or the other is not registered properly, the designs will not match up on both sides after the sheets are cut up; if the registration is very poor, the notes will be replaced. A third separate printing operation adds the serial number and Treasury Seal; the major error possibilities are in inking and placement, since the basic note design already exists at this point.

You have probably noticed that serial numbers on U.S. currency are preceded by a letter. That letter designates which of the twelve Federal Reserve districts issued the note (this is why the letters span A through L). For example, all serial numbers preceded by D (the fourth letter of the alphabet) are issued by the Fourth District of the Federal Reserve (Cleveland). Here is a list of the twelve Federal Reserve Bank districts and the letter designations for each:

District	Letter	City
1	A	Boston
2	B	New York
3	C	Philadelphia
4	D	Cleveland
5	E	Richmond
6	F	Atlanta
7	G	Chicago
8	H	St. Louis
9	I	Minneapolis
10	J	Kansas City
11	K	Dallas
12	L	San Francisco

The star note enables the treasury to issue a new set of serial numbers rather than attempting to reassign all the missing serial numbers of defective notes. On U.S. notes, a star substitutes for

WHEN DO FISH SLEEP? 181

the prefix letter. A replacement U.S. note might look like this: ★ 00000007 B. On Federal Reserve notes, a star substitutes for the letter at the end of the serial number, so that the location of the Federal Reserve district is kept intact: D 00000007 ★.

William Bischoff, associate curator of the American Numismatic Society, adds that there is one other use for the star note. The Bureau of Printing and Engraving uses printers with eight-digit numbering cylinders to produce one hundred million notes at a time. But for the one-hundred-millionth note, a ninth digit is needed. Rather than bothering to add another digit on the cylinder that would literally be used on one out of a hundred million notes, the one-hundred-millionth note is a hand-inserted star note.

To What Do the Numbers Assigned to Automotive Oil Refer?

Thirty years ago, 10–30 was considered a premium automotive oil. Today, one can buy 10–50 or even 10–60 oil, but few people know what these numbers mean.

The numbers measure the viscosity of the oil. The higher the number, the higher the viscosity (meaning the oil is less likely to flow). Although the viscosity of a liquid is not always directly correlated to thickness, high-viscosity oils are thicker than their low-viscosity counterparts.

The numbers on engine and transmission oils are assigned by the Society of Automotive Engineers. Their numbers range from 5W to 60. The W stands for winter. When a W follows a number, it indicates the viscosity of the oil at a low temperature. When there is no W following a number, the viscosity is measured at a high temperature.

All oil companies promote multigrade oils, which are designed to perform well in hot or cold temperatures. Thus 10W–40 doesn't indicate a range of viscosity, but rather the low viscosity of oil during winter (when one desires greater flow capabilities) and high viscosity in the summer.

Submitted by Tom and Marcia Bova of Rochester, New York.

When I Put One Slice of Bread in My Toaster, the Heating Element in the Adjacent Slot Heats Up as Well. So Why Does My Toaster Specify Which Slot to Place the Bread in If I Am Toasting Only One Slice?

Considering that the pop-up toaster has proven to be perhaps the most durable and dependable kitchen appliance, we were surprised to learn that toasting technology varies considerably from model to model. The earliest toasters browned one side of bread at a time; one had to decide when to flip the bread over by hand, a problem not unlike the momentous decision of when to flip over a frying pancake or hamburger.

Now that even the simplest pop-up toaster has a toast selector dial to allow the user to choose the preferred degree of doneness, most of the guesswork in toasting has been eliminated. We are not even allowed to select which of two or four slots to put in our one meager slice of bread. Why not?

Actually, nothing dire will result if you don't use the slot marked ONE SLICE. The worst that will happen is that the toaster will pop up an underdone or overdone piece. But why is the same well that manages to produce wonderful toast when it has company next door suddenly rendered incompetent when forced to work alone?

The answer depends upon the type of technology the toaster uses to determine doneness. The simplest toasters, now passé, worked from a simple time principle. The darker the brownness dial was set for, the longer the timer set for the toaster to heat the bread. Toasters that worked on a timer alone did not need a ONE SLICE notation because they always cooked the bread for the same amount of time, as long as the brownness dial wasn't changed. Using a timer alone guaranteed that a second set of toast would come out overdone, because the toaster was already warmed up yet toasted the second set for the same period of time as the first batch that was heated from a "cold start."

To solve the problem, appliancemakers inserted a thermostatic switch in toasters, which measured the heat of the toaster rather than the time elapsed in cooking. The thermostat alone caused a reverse problem. A second batch of bread would come out underdone because the first cycle had already caused the heating element to charge. The toaster didn't "know" that the second batch of bread hadn't been exposed to the toaster long enough; it knew only that the toaster had achieved the desired temperature.

The solution to the problem was to use a combination timer-thermostat. Today, the timer is not set off until the thermostat tells the timer that the toaster has reached the preset temperature (determined by the setting of the brownness dial). With this technology, it might take a minute for the thermostat to tell the timer to start ticking with the first set of toast but only a few seconds for the second or third.

We spoke to an engineer at Proctor-Silex who told us that most of their toasters have the thermostat close to—or in some cases, inside—the well that is marked ONE SLICE so the thermostat can do a more accurate job of "reading" the correct temperature for that slice. Some toasters that have ONE SLICE markings are "energy saver" toasters, specifically designed so that the heating element in the second slot will not be charged if it does not contain bread.

Sunbeam has long produced the 20030 toaster, an elegant two-slicer that selects the proper brownness of the bread by a radiant control that "reads" the surface of the bread to determine the degree of doneness. As far as we know, the Sunbeam 20030 is the only toaster that doesn't work on a time principle. The 20030 actually measures the surface temperature of the bread by determining its moisture level and can accurately measure the time needed to toast any type of bread. Wayne R. Smith, of

　　　　　　　　　　　　　　DAVID FELDMAN

Sunbeam Public Relations, told *Imponderables,* "There's no point in having radiant controls in both slots when having a control in one slot works just as well."

Submitted by Lisa M. Giordano of Tenafly, New Jersey. Thanks also to Muriel S. Marschke of Katonah, New York; and Jim Francis of Seattle, Washington.

Why Are Almost All Cameras Black?

Black isn't the most obvious color we would pick for cameras. Not only is black an austere and a threateningly high-tech color to amateurs, it would seem to have a practical disadvantage. As Jim Zuckerman, of Associated Photographers International, explained, black tends to absorb heat more than lighter colors, and heat is the enemy of film.

Of course, there was and is no reason why the exteriors of cameras need to be black. For a while, chromium finishes were popular on 35 millimeter cameras, but professional photographers put black tape over the finish to kill any possible reflections. Sure, some companies now market inexpensive cameras with decorator colors on the exterior. Truth be told, the persistence of black exteriors on cameras has more to do with marketing than anything else. As Tom Dufficy, of the National Association of Photographic Manufacturers, told us: "To the public, black equals professional."

Submitted by Herbert Kraut of Forest Hills, New York.

WELL, THEY SAID MY NEW SHIRT WOULD IRON ITSELF!!

SHIRT PERMA PRESS #39.95

Why Is There a Permanent Press Setting on Irons?

We buy a permanent press shirt so that we won't have to iron it. Then after we wash the shirt for the first time, it comes out of the dryer with wrinkles. Disgusted, we pull out our iron only to find that it has a permanent-press setting. Are iron manufacturers bribing clothiers to renege on their promises? Is this a Communist plot?

The appliance industry is evidently willing to acknowledge what the clothing industry is reluctant to admit: A garment is usually permanently pressed only until you've worn it—once. Wayne R. Smith, consultant in Public Relations to the Sunbeam Appliance Company, suggested that "permanent press" was chosen to describe the benefits of some synthetic materials be-

DAVID FELDMAN

cause "it has a far more attractive sound to consumers than 'wrinkle-resistant.' "

We know what Mr. Smith means. We've always felt that the difference between a water-resistant watch and a waterproof watch was that the waterproof one would die the moment *after* it hit H_2O.

What Causes Double-Yolk Eggs? Why Do Egg Yolks Sometimes Have Red Spots on Them?

Female chicks are born with a fully formed ovary containing several thousand tiny ova, which form in a cluster like grapes. A follicle-stimulating hormone in the bloodstream develops these ova, which will eventually become egg yolks. When the ova are ripe, the follicle ruptures and an ovum is released. Usually when a chicken ovulates, one yolk at a time is released and travels down the oviduct, where it will acquire a surrounding white membrane and shell.

But occasionally two yolks are released at the same time. Double-yolk eggs are no more planned than human twins. But some chickens are more likely to lay double-yolk eggs. Very young and very old chickens are most likely to lay double yolks; young ones because they don't have their laying cycles synchronized, and old ones because, generally speaking, the older the chicken, the larger the egg she will lay. And for some reason, extra-large and jumbo eggs are most subject to double yolks.

If a chicken is startled during egg formation, small blood vessels in the wall may rupture, producing in the yolk blood spots—tiny flecks of blood. Most eggs with blood spots are removed when eggs are graded, although they are perfectly safe to eat.

DAVID FELDMAN

Submitted by Lewis Conn of San Jose, California. Thanks also to Melody L. Love of Denver, North Carolina.

Why Are Barns Red?

We first encountered this Imponderable when a listener of Jim Eason's marvelous KGO-San Francisco radio show posed it. "Ummmmm," we stuttered.

Soon we were bombarded with theories. One caller insisted that red absorbed heat well, certainly an advantage when barns had no heating system. Talk-show host and guest agreed it made some sense, but didn't quite buy it. Wouldn't other colors absorb more heat? Why didn't they paint barns black instead?

Then letters from the Bay area started coming in. Donna Nadimi theorized that cows had trouble discriminating between different colors and just as a bull notices the matador's cape, so a red barn attracts the notice of cows. She added: "I come from West Virginia and once asked a farmer this question. He told me that cows aren't very smart, and because the color red stands out to them, it helps them find their way home." The problem with this theory is that bulls are color-blind. It is the movement of the cape, not the color, that provokes them.

Another writer suggested that red would be more visible to owners, as well as animals, in a snowstorm. Plausible, but a stretch.

Another Jim Eason fan, Kemper "K.C." Stone, had some "suspicions" about an answer. Actually, he was right on the mark:

> The fact is that red pigment is cheap and readily available from natural sources. Iron oxide—rust—is what makes brick clay the color that it is. That's the shade of red that we westerners are accustomed to—the rusty red we use to stain our redwood decks. It's obviously fairly stable too, since rust can't rust and ain't likely to fade.

The combination of cheapness and easy availability made red an almost inevitable choice. Shari Hiller, a color specialist at the Sherwin-Williams Company, says that many modern barns are painted a brighter red than in earlier times for aesthetic reasons. But aesthetics was not the first thing on the mind of farmers painting barns, as Ms. Hiller explains:

> You may have noticed that older barns are the true "barn red." It is a very earthy brownish-red color. Unlike some of the more vibrant reds of today that are chosen for their decorative value, true barn red was selected for cost and protection. When a barn was built, it was built to last. The time and expense of it was monumental to a farmer. This huge wooden structure needed to be protected as economically as possible. The least expensive paint pigments were those that came from the earth.

Farmers mixed their own paint from ingredients that were readily available, combining iron oxide with skim milk—did they call the shade "2% red"?—linseed oil and lime. Jerry Rafats, reference librarian at the National Agricultural Library, adds that white and colored hiding pigments are usually the most costly ingredients in paints.

K.C. speculated that white, the most popular color for buildings in the eighteenth and nineteenth centuries (see *Why Do Clocks Run Clockwise? and Other Imponderables* for more than you want to know about why most homes are and always have been painted white), was unacceptable to farmers because it required constant cleaning and touching up to retain its charm. And we'd like to think that just maybe the farmers got a kick out of having a red barn. As K.C. said, "Red is eye-catching and looks good, whether it's on a barn, a fire truck, or a Corvette."

Submitted by Kemper "K.C." Stone of Sacramento, California. Thanks also to Donna Nadimi of El Sobrante, California; Jim Eason of San Francisco, California; Raymond Gohring of Pepper Pike, Ohio; Stephanie Snow of Webster, New York; and Bettina Nyman of Winnipeg, Manitoba.

DAVID FELDMAN

Why Are Manhole Covers Round?

On one momentous day we were sitting at home, pondering the imponderable, when the phone rang.

"Hello," we said wittily.

"Hi. Are you the guy who answers stupid questions for a living?" asked the penetrating voice of a woman who later introduced herself as Helen Schwager, a friend of a friend.

"That's our business, all right."

"Then I have a stupid question for you. Why are manhole covers round?"

Much to Helen's surprise, the issue of round manhole covers had never been important to us.

"Dunno."

"Guess!" she challenged.

So we guessed. Our first theory was that a round shape roughly approximated the human form. And a circle big enough to allow a worker would take up less space than a rectangle.

"Nope," said Helen, friend of our soon-to-be ex-friend. "Try again."

Brainstorming, a second brilliant speculation passed our lips. "It's round so they can roll the manhole cover. Try rolling a heavy rectangular or trapezoidal manhole cover on the street."

"Be serious," Helen insisted.

"O.K., we give up. Tell us, O brilliant Helen. *Why are manhole covers round?*"

"It's obvious, isn't it?" gloated Helen, virtually flooding with condescension. "If a manhole were a square or a rectangle, the cover could fall into the hole when turned diagonally on its edge."

Helen, who was starting to get on our nerves just a tad, went on to regale us with the story of how she was presented with this Imponderable at a business meeting and came up with the answer on the spot. With tail between our legs, we got off the phone, mumbling something about maybe this Imponderable getting in the next book. First we get humiliated by this woman; then we have to give her a free book. Isn't there any justice?

Of course, after disconnecting with Helen we did what any self-respecting American would do: We tortured our friends with this Imponderable, making them feel like pieces of dogmeat if they didn't get the correct answer. And very few did.

Of course, we can't rely on an answer provided by the supplier of an Imponderable, even one so intelligent as Helen, so we contacted many manufacturers of manhole covers, as well as city sewer departments.

Guess what? The manufacturers of manhole covers can't agree on why manhole covers are round. Some, such as the Vulcan Foundry of Denham Springs, Louisiana, immediately confirmed Helen's answer but couldn't resist throwing a plug in as well ("Then, again, maybe manhole covers are round to facilitate the use of the Vulcan Classic Cover Collection").

But the majority of the companies we spoke to said not only do manhole covers not have to be round but many aren't. Manhole covers sit inside a frame or a ring that is laid into the concrete. Many of these frames cover the hole completely and are not hollow, so there is no way that a cover any shape could fall into the hole.

Most important, as Eric Butterfield, of Emhart Corporation, told *Imponderables*, manhole covers have a lip. Usually the manhole cover is at least one inch longer in diameter for each foot of the diameter of the hole.

Round manholes are more convenient in other ways. Lathe workers find circular products easier to manufacture. Seals tend to be tighter on round covers. And Lois Hertzman, of OPW, a division of Dover Corporation, adds that round manholes are easier to install because there are no edges to square off.

Everyone we spoke to mentioned that many manholes are not round. Many older manhole covers are rectangular. The American Petroleum Institute wants oil covers to be the shape of equilateral triangles (impractical on roadways, where this shape could lead to covers flipping over like tiddlywinks).

Engineers at the New York City Sewer Design Department

could find no technical reason for round manhole covers. They assumed, like most of the fall-through theory dissenters, that the round shape is the result of custom and standardization rather than necessity.

So, Helen, we have wreaked our revenge. Perhaps your answer is correct. But then, maybe it is wrong. Maybe the real reason manholes are round is so that they can facilitate the use of the Vulcan Classic Cover Collection.

Submitted by Helen Schwager of New York, New York.
Thanks also to Tracie Ramsey of Portsmouth, Virginia; and
Charles Kluepfel of Bloomfield, New Jersey.

Frustables

OR

The 10 Most Wanted Imponderables

In *Why Do Clocks Run Clockwise?*, we broke down and admitted we were haunted and rendered sleepless by our inability to answer some Imponderables that were sent to us. Often we found fascinating explanations, tantalizing theories, or partial proof. But burdened by the strict ethical codes that the custody of the body Imponderability places upon us, we can't rest until we positively nail the answers to these suckers.

So we asked our readers for help with the ten most Frustrating Imponderables (or Frustables, for short). The fruits of your labors are contained in the following pages. But before you get totally smug about your accomplishments, may we lay ten more on you?

These are ten Imponderables for which we don't yet have a conclusive answer. Can you help? A reward of a free, autographed copy of the next volume of *Imponderables,* as well as an acknowledgment in the book, will be given to the first reader who can lead to the proof that solves any of these Frustables.

FRUSTABLE 1: *Why is Legal Paper 8½″ × 14″?*

We have located the first company to manufacture a legal-sized pad. We've also contacted the largest manufacturers of paper and stationery and many legal sources. But no one seems to know the reasons for lengthening regular paper and dubbing it "legal size." And yes, we know that many courts have abandoned legal-sized paper and now use 8½″ × 11″.

FRUSTABLE 2: *Why Do Americans, Unlike Europeans, Switch Forks to the Right Hand After Cutting Meat?*

Did someone give the Pilgrims radical etiquette lessons on the *Mayflower?* Is there any sense to the American method?

FRUSTABLE 3: *How, When, and Why Did the Banana Peel Become the Universal Slipping Agent in Vaudeville and Movies?*

Vegetable oil would work better, no?

FRUSTABLE 4: *Why did the Grade E Disappear from Grading Scales in Most Schools?*

An F makes sense as the lowest mark (F = failure); but why did the E get lost?

FRUSTABLE 5: *How Did They Lock Saloon Doors in the Old West?*

Were saloons in the old West open 24 hours? If they weren't, a couple of swinging doors three feet off the ground wouldn't provide a heckuva lot of security. Were there barriers that covered the entrance, or are swinging saloon doors a figment of movemakers' imaginations?

DAVID FELDMAN

FRUSTABLE 6: *Why Do So Many People Save* National Geographics *and Then Never Look at Them Again?*

A visit to just about any garage sale will confirm that most people save *National Geographics*. An unscientific poll confirms that nobody ever looks at the issues they've saved. What gives?

FRUSTABLE 7: *Why Do People, Especially Kids, Tend to Stick Their Tongues Out When Concentrating?*

Theories abound, but no one we contacted had any confidence about their conjectures.

FRUSTABLE 8: *Why Do Kids Tend to Like Meat Well Done (and Then Prefer It Rarer and Rarer as They Get Older)?*

Are kids repelled by the sight of blood in rare meat? Do they dislike the texture? The purer taste of meat? What accounts for the change as they get older?

FRUSTABLE 9: *Why Does Whistling at an American Sporting Event Mean "Yay!" When Whistling Means "Boo!" in Most Other Countries?*

FRUSTABLE 10: *Why Are So Many Restaurants, Especially Diners and Coffee Shops, Obsessed with Mating Ketchup Bottles at the End of the Day?*

We have been in sleazy diners where we couldn't hail a waitress if our lives depended on it and were lucky if our table was cleaned off. Where was the waitress? She was grabbing all the ketchup bottles and stacking them so that the remains of one bottle flowed into a second bottle.

Who cares whether the ketchup bottle on the table is one third filled or completely full? Doesn't the ketchup flow more easily out of a less than full bottle? Do restaurateurs mate ketchup bottles to please patrons, or do they have other, perhaps nefarious, reasons?

Frustables Update

Captured!!! The Ten Frustables from Why Do Clocks Run Clockwise?

FRUSTABLE 1: *Why Do You So Often See One Shoe Lying on the Side of the Road?*

They say that every parent has a favorite. In this case, we'll admit it. This isn't only our favorite Frustable, it's probably our favorite Imponderable ever, partly because it has been a difficult "child." We spoke to endless officials at the Department of Transportation and the Federal Highway Safety Traffic Administration. All of them were aware of the phenomenon; none had a compelling explanation.

In *Why Do Clocks Run Clockwise?* we talked about some of

the theories proffered by readers of Elaine Viets, columnist for the *St. Louis Post-Dispatch:*

- They are tossed out of cars during fights among kids.
- They fall out of garbage trucks.
- Both shoes are abandoned at the same time, but one rolls away.
- They are disentangled, discarded newlywed shoes.
- They are thrown out of school buses and cars as practical jokes.

We asked if our readers could come up with anything better.

We needn't have worried. You guys came through in spades. Your answers fell into three general categories: theoretical, empirical, and confessional. So profound were your insights into this important subject that we have given thirteen of them official *Imponderables* Awards of Merit.

Award-Winning Theoretical Explanations

Best Supply-Side Argument by a Noneconomist. Provided by Stefan Habsburg of Farmington Hills, Michigan: "Because if there were a pair, someone would pick them up!"

Best Conspiracy Theory. Provided by Morry Markovitz of Croton Falls, New York:

> If a lost pair of shoes were found intact, the shoe industry might lose a sale as these old shoes were pressed into service by a new owner. Has the shoe industry secretly hired "road agents" to scour the country-side, picking up one of each pair they find?

Best Explanation Involving Eastern European Influence Upon the One-Shoe Problem. Provided by Rick La Komp of Livermore, California: "Barefoot field-goal kickers decided they didn't need more than one shoe and threw the other away."

Best Explanation by an Obnoxious Anthropomorphic Cartoon Animal. Provided by David Selzler of Loveland, Colorado. David sent us a "Garfield" cartoon in which the cat muses, "Why do you find only one shoe in the trash? One shoe on a sidewalk? One shoe in the street?" He wonders about why people don't throw things away in pairs. So Garfield sees one shoe in a trash can and knocks on the door of the adjacent house. Guess who answers? A pirate with a peg leg.

Most Logical Theory. Provided by Russ Tremayne of Auburn, Washington, and Maria N. Benninghoven of Kensington, Maryland. Both of these readers assume that most shoes found on the side of the road are thrown out of moving cars. They also assume that most people toss both shoes out one at a time. Russ assumes it would be most natural to throw out the shoes with one dominant hand:

> Most people's hands aren't large enough to comfortably grasp a pair of shoes, even if the laces are tied. Therefore, one shoe gets thrown at a time as the vehicle continues to travel. Perhaps one shoe, thrown weakly, lands on the edge of the highway, while the other, thrown with more force, lands off the road to lie invisibly among tall grass or brush.

Empirical Theories

Best Exploitation for Personal Profit of the One-Shoe Phenomenon. Provided by R.E. Holtslander of Lake Wales, Florida.

> About 20 years ago when I lived in Missouri and was coming home from California on a windy day, I noticed a large cardboard box on the highway in New Mexico. Papers were flying from it. Shortly after that, I saw an almost new broom, so I pulled off the road and picked it up.
>
> Soon other things appeared by the road. I saw a shoe for the right foot. As I had a sore toe at the time and thought the shoe was big enough to give my foot comfort, I picked it up, too.
>
> I saw a man in a pickup truck at the side of the road. He too had stopped to retrieve something. From then on it was like a treasure hunt. I picked up several things and then he would pass us and then we would pass him. Soon we passed into Texas, and there I found the mate to the shoe I had picked up in New Mexico!
>
> I kept the shoes for several years and showed my guests a pair of shoes, one of which I got in New Mexico and the other in Texas . . .

Best Explanation for the Unsalutary Effect of Poor Nutrition and Sleeping Habits Upon the Retention of Shoes. Provided by Dave Sodovy of Hamilton, New York: Dave recounts the story that a few summers ago he had a job with two other kids who were the sons of the boss. Father and sons lived an hour's drive away from work, necessitating leaving their house at 7:15 because the boss wanted to get coffee and doughnuts to fortify him for the road. The sons retaliated for having to get up at this barbaric hour by sleeping through the trip.

From this evolved a routine in which the time between waking up and later falling asleep in the car was spent in "semi-sleep."

One morning, the two boys and their dad arrived as usual, but the younger son was wearing only one shoe! A few questions revealed the reason. In his state of semi-sleep, with one shoe on, one shoe in one hand, and a bag lunch in the other hand, he set the shoe on the roof of the car to open the car door.

Since his main concern was to go to sleep in dad's car, he didn't retrieve his shoe—he just closed the door and got comfortable. Dad and his other son had been in the car waiting; the three took off as soon as my one-shoed friend closed the door. The shoe was still on the roof of the car, and apparently survived the 25- and 35-mph speed limits of the neighborhood in which they lived. Once on the highway, the shoe was doomed. Indeed, once the three arrived at work, they called Mom, who sought out the missing shoe, locating it on the side of the highway.

I doubt that this is the definitive answer that you're looking for, but it does explain how at least one shoe got on the road by itself.

The Margaret Mead Field Research Award. Goes to Laurie McDonald of Houston, Texas. Laurie once lived in Providence, Rhode Island, and found scores of single shoes alongside the highway between Warwick and Pawtucket while driving to and from work. Laurie collected seventy shoes in a period of six months.

Among her discoveries were mostly tennis shoes, most left-footed; "very early on Sunday mornings, I would inevitably find a few brand-new patent leather platform shoes sitting upright on the side of the road, waiting to be plucked."

In 1979, Laurie was inspired to write a series of short stories, consisting of hypothetical explanations of how single shoes landed off the highway. In one, the protagonist was a hippie, sticking out his hand to hail a car. Instead, the outstretched hand became a shooting target for unneeded boots of army servicemen.

Although Laurie offers no theoretical breakthroughs, we nevertheless owe her a great deal for proving conclusively that single shoes are deposited on the highway at an alarming rate, at least in Rhode Island.

Confessions

The "10–4, Good Buddy" Award. Goes to Robin Barlett of Erie, Pennsylvania. Robin writes:

> My husband is a truck driver and I went with him on a run. He got tired and pulled off to the side of the road to sleep, and I went back into the bunk and took my shoes off.
>
> We got up late at night and my husband had to go to the bathroom. Nobody was on the highway so he just hopped out to go and he must have kicked my shoe out, for it was not there when I tried to find it.

The Foot Out the Window Routine Award. Goes to Brian Razen Cain of Chipley, Florida, the last honest man in the world. Many readers theorized that single shoes are remnants of passengers who nap with one (or both) shoes dangling out car windows. But Brian was the only one to admit it. Brian fell asleep on the Florida Turnpike and woke up to find himself semishoeless. His response? The normal one: "I simply threw the other one out the window too."

The Dog Ate My Homework Award. Goes to Jay Lewis of Montgomery, Alabama.

> The shoes were dropped there by animals. Various beasties are attracted by the taste and smell of salt-impregnated leather. Since animals have trouble getting more than one shoe in their mouths,

they only carry one of the pair away. Where they finally discard it is where you see it—invariably without its mate.

C. Lynn Graham of Pelham, Alabama, adds that dogs tend to think that they will carry a shoe in their mouths forever. But as soon as they are distracted by a car coming down the road, the shoe pales compared to the chance to chase a passing car.

Surprisingly only one soul, Jennifer Ballmann of Jemez Springs, New Mexico, was willing to admit that she was the personal victim of doggy single-shoe syndrome. Two questions arise: Can a Saint Bernard carry two shoes at a time? Can a Pekingese carry one shoe at a time?

I Did it for the Sake of Science Award. Goes to an anonymous caller on a Detroit, Michigan, radio show hosted by David Newman. This caller, a paramedic, confessed that he was personally responsible for dropping several single shoes in the past week.

When administering CPR paramedics are trained to take off the victim's shoes, in order to promote better circulation. Many times in transporting a heart attack victim from a residence the shoes are taken off hastily and get lost before the patient enters the ambulance.

Will insurance companies pay for the missing shoes of patients not responsible for their loss?

The Motorcycles are Dangerous Even WITH a Helmet Award. Goes to Tom Vencuss of Newburgh, New York. Several readers speculated that single shoes were discarded wedding shoes. "Well," claims Tom, "not quite . . . "

> I was scheduled to be the best man at my cousin's wedding in Schenectady, New York, a two-hour ride from my home in Poughkeepsie. Prior to the wedding, I needed to make a quick run up to get fitted for my tuxedo. It was a beautiful afternoon so I decided to ride my motorcycle. I dressed in my normal riding gear but took a pair of dress shoes to wear at the fitting. I decided not to take a bag along since I would not be spending the night. So I strapped the shoes to the back of the bike.

Several hours later, as I pulled into the parking lot for the tuxedo rental, I reached back to find only one shoe. The other, no doubt, was sitting lonely on a stretch of the New York State Thruway. Though it was an expensive afternoon, it did solve one of life's little mysteries.

Do we have the definitive, smoking-gun solution to this Frustable? We're afraid not. But after all, philosophers have been arguing over less important topics for thousands of years. Together we have raised the level of discourse on this topic to stratospheric heights. Maybe our grandchildren will find the ultimate answer.

Submitted by Julie Mercer of Baltimore, Maryland. Thanks also to Bess M. Bloom of Issaquah, Washington; and Sue S. Child of Red Bluff, Louisiana.

A free book goes to Laurie McDonald of Houston, Texas, to inspire her to conduct further hard research on this important topic. Thanks also to Elaine Viets of the St. Louis Post-Dispatch, *for her generosity.*

FRUSTABLE 2: *Why Are Buttons on Men's Shirts and Jackets Arranged Differently From Those on Women's Shirts?*

Of all the Frustables, none yielded less new ground than this one. Although more readers tried to answer this Frustable than any other but number three, few added much beyond the speculations we offered in *Why Do Clocks Run Clockwise?*

This much we know for sure: Buttons were popularized in the thirteenth century, probably in France. Before that, robes tended to be loose and unfitted and were fastened by strings, hooks, or pins.

Many readers said that men in the thirteenth century wore swords on the left hip under their coats. When they cross-drew their sword, they risked catching their sword if their garments were arranged right over left. By changing the configuration so that left closed over right, they could unfasten jacket buttons with their left hand and draw their sword with their right hand more quickly and safely.

An equal number of readers insisted that the different button configurations come from the custom of rich women having handmaidens who dressed them. Because clothes, as everything else, were designed for right-handers, the women's arrangement made it easier for maids to button their mistresses' blouses and dresses. Male aristocrats presumably dressed themselves.

The third popular answer is that women's button arrangement is most convenient for breast-feeding, so that the mother can unbutton her blouse with her right hand and rest the baby on her left arm.

None of the three popular explanations is convincing to us. All of the clothing historians we spoke to did not accept these pat answers either. Only the second theory explains why men's and women's buttons need to be different, and there is one inherent problem with this theory—many rich men were indeed dressed by servants. We don't put much stock in the breast-feeding theory either. One book we found mentioned that the women's arrangement made it easier for women who were breast-feeding while holding their babies in their right arms.

Much of the written material we have read on this subject mentions all three of these nonrelated theories, an indication to us that these explanations are based more on supposition after the fact than solid evidence. Robert Kaufman, reference librarian in the Costume Division at the Metropolitan Museum of Art in New York, told *Imponderables* that this Frustable has been among his most often asked questions. He and others have done considerable research on the subject and found no credible evidence to sustain any particular argument.

DAVID FELDMAN

A few readers offered some imaginative answers to this Frustable. A surprising number mentioned an intriguing variation of the handmaiden theory: By switching the button arrangement for the two sexes, it's easier for the two sexes to unbutton each other's clothing during a sexual encounter. Hmmmmmm.

Along the same garden path comes our favorite contribution to this discussion, from Erik Johnson of Houston, Texas: "The button arrangement is so that when a couple is driving in a car, with the man driving, they can peek inside each other's shirts."

Submitted by Julia Zumba, of Ocala, Florida. Thanks also to Kathi Sawyer-Young of Encino, California; Mathew Gradet of Ocean City, Maryland; Jodi Harrison of Helena, Montana; Sheryl K. Prien of Sacramento, California; Harry Geller of Rockaway Beach, New York; Terry L. Stibal of Belleville, Illinois; Mary Jo Hildyard of West Bend, Wisconsin; Tom and Marcia Bova of Rochester, New York; Robert Hittel of Fort Lauderdale, Florida; and many others.

FRUSTABLE 3: *Why Do the English Drive on the Left and Just About Everybody Else on the Right?*

Quite a few readers, justifiably, took us to task for the phrasing of this Frustable. By saying "just about everybody else" drives on the right, we didn't mean to slight the rest of the United Kingdom, Ireland, India, Indonesia, Australia, South Africa, Kenya, Thailand, Japan, and many other nations, all of which still drive on the left side of the road. But because most of these other countries adopted their traditions while a part of the British Empire, we wanted to give the "credit" where it was due.

So how did this left-right division start? All historical evidence indicates that in ancient times, when roads were usually narrow and unpaved, a traveler would move to the left when encountering another person on foot or horse coming toward them. This allowed both parties to draw their sword with their right hand, if necessary. If the approaching person were friendly, one could give the other a high five instead. Military policy, as far back as the ancient Greeks, dictated staying to the left if traveling without a shield, so that a combatant could use his left hand to hold the reins and need not brandish the sword or lance crosswise, risking the neck of the horse.

Richard H. Hopper, a retired geologist for Caltex, has written a wonderful article, "Why Driving Rules Differ," the contents of which he was kind enough to share with *Imponderables* readers. Hopper believes that the custom of mounting horses on the left-hand side also contributed to traffic bearing left. In many countries, pedestals were placed alongside the curbs of the road to help riders mount and dismount from their horses. These approximately three-feet-high pedestals were found only on the left side of the road. Long before the pedestals were erected, horsemen mounted and dismounted on the left, probably because their scabbards, slung on the left, interfered with mounting the horse; the unencumbered right leg could be more easily lifted over the horse.

Until 1300 A.D., no nation had mandated traffic flow. But Pope Boniface VIII, who declared "all roads lead to Rome," insisted that all pilgrims to Rome must stick on the left side of the road. According to Hopper, "This edict had something of the force of law in much of western Europe for over 500 years."

The movers and shakers of the French Revolution weren't excited about having a pope dictate their traffic regulations. Robespierre and other Jacobins encouraged France to switch to right-hand driving. Napoleon institutionalized the switch, not only in France but in all countries conquered by France.

Why did the United States, with an English heritage, adopt the French style? The answer, according to Hopper and many others, is that the design of late-eighteenth-century freight wagons encouraged right-hand driving. Most American freight wagons were drawn by six or eight horses hitched in pairs; the most famous of these were the Conestoga wagons that hauled wheat from the Conestoga Valley of Pennsylvania to nearby cities. These wagons had no driver's seat. The driver sat on the left-rear horse, holding a whip in his right hand. When passing another vehicle on a narrow ride, the driver naturally went to the right, to make sure that he could see that the left axle hub and wheel of his wagon were not going to touch those of the approaching vehicle.

In 1792, Pennsylvania passed the first law in the United States requiring driving on the right-hand side, although this ordinance referred only to the turnpike between Lancaster and Philadelphia. Within twenty years, many more states passed similar measures. Logically, early American carmakers put steering wheels on the left, so that drivers on two-lane roads could evade wavering oncoming traffic. Although Canada originally began driving on the left-hand side, the manufacture of automobiles by their neighbors to the south inevitably led to their switch to the right. Although Ontario adopted right-hand driving in 1812, many other provinces didn't relent until the 1920s.

Great Britain, of course, stayed with the ancient tradition of left-side driving and not just out of spite. Their freight wagons were smaller than Conestoga wagons and contained a driver's seat. Hopper explains why the driver sat on the right-hand side of the wagon:

> The driver sat on the right side of the seat so that he could wield his long whip in his right hand without interference from the load behind him. In passing oncoming wagons, the drivers tended to keep to the left of the road, again to be able to pass approaching vehicles as closely as necessary without hitting.
>
> In passenger carriages, the driver also sat on the right, and the footman, if there was one, sat to the driver's left so that he could quickly jump down and help the passengers disembark at the curb.

Needless to say, a coachman wouldn't have felt quite as secure sitting to the right of the driver. Every time a right-handed driver got ready to crack the whip, the coachman would have had to duck and cover.

The British built their cars with the steering wheel on the right because their wagons and carriages at the time still stuck to the left side of the road. Their foot controls, however, have always been the same as American cars.

Well more than a hundred readers sent responses to this Frustable, most of them containing fragments of this explanation. Hopper's article is the best summary of the conventional wisdom on this subject that we have encountered. But there are dissenters. Patricia A. Guy, a reference librarian at the Bay Area Library and Information System in Oakland, California, was kind enough to send us several articles on this subject, including a fascinating one called "The Rule of the Road" from a 1908 periodical called *Popular Science Monthly*. Its author, George M. Gould, M.D., argues that Americans had adopted right-hand side travel before the development of Conestoga wagons, as had the French, whose wagons were driven by postilion riders (mounted on the left-rear horse). Dr. Gould couldn't come up with a convincing theory for the switch and argued that this Imponderable was likely to be a Frustable for all time.

We include this dissent to indicate that we tend to lunge at any answer that neatly solves a difficult question. We can give you a logical reason why Americans and the French switched the traditional custom of driving on the left; but we wouldn't risk our already precarious reputations on it.

Submitted by Claudia Wiehl of North Charleroi, Pennsylvania. Thanks also to John Haynes of Independence, Kentucky; Kathi Sawyer-Young of Encino, California; Larry S. Londre of Studio City, California; David Andrews of Dallas, Texas; Hugo Kahn of New York, New York; Barbara Dilworth of Bloomsburg, Pennsylvania; Pat Mooney of Inglewood, California; Stephen Murphy of Smithfield, North Carolina; Frederick A. Fink of Coronado, California; and many others.

A free book goes to Richard H. Hopper of Fairfield, Connecticut.

DAVID FELDMAN

FRUSTABLE 4: *Why Is Yawning Contagious?*

After the publication of *Imponderables*, this question quickly became our most frequently asked Imponderable. And after years of research, it became one of our most nagging Frustables. We couldn't find anyone who studied yawning, so we asked our readers for help.

As usual, our readers were bursting with answers, unfortunately, conflicting answers. They fell into three classes.

The Physiological Theory. Proponents of this theory stated that science has proven that we yawn to get more oxygen into our system or to rid ourselves of excess carbon dioxide. Yawning is contagious because everybody in any given room is likely to be short of fresh air at the same time.

The Boredom Theory. If everyone hears a boring speech, why shouldn't everyone yawn at approximately the same time, wonders this group.

The Evolutionary Theory. Many readers analogized contagious yawning in humans to animals displaying their teeth as a sign of intimidation and territoriality. Larry Rose of Kalamazoo, Michigan, argued that yawning might have originally been a challenge to others, but has lost its fangs as an aggressive maneuver as we have gotten more "civilized."

Several readers pointed us in the direction of Dr. Robert Provine, of the University of Maryland at Baltimore County, who somehow had eluded us. You can imagine our excitement when we learned that Dr. Provine, a psychologist specializing in psychobiology, is not only the world's foremost authority on yawning, but has a special interest in why yawning is contagious! In one fell swoop, we had found someone who not only might be able to answer a Frustable but a fellow researcher whose work was almost as weird as ours.

Dr. Provine turned out to be an exceedingly interesting and generous source, and all the material below is a distillation of his work. As usual, experts are much less likely to profess certainty about answers to Imponderables than are laymen. In fact, Provine confesses that we still don't know much about yawning; what we do know is in large part due to his research.

Provine defines yawning as the gaping of the mouth accompanied by a long inspiration followed by a shorter expiration. This definition seems to support the thinking of some who believe the purpose of a yawn is to draw more oxygen into the system, but Provine disagrees. He conducted an experiment in which he taped the mouths of his subjects shut. Although they could yawn without opening their mouths, they felt unsatisfied, as if they weren't really yawning, even though their noses were clear and were capable of drawing in as much oxygen as if their mouths were open. From this experiment, Provine concludes that the function of yawning is not related to respiration.

In other experiments, Provine has proven that yawning has nothing to do with oxygen or carbon dioxide intake. When he pumped pure oxygen into subjects, for example, their frequency of yawning did not change.

DAVID FELDMAN

Provine's research also supports the relationship between boredom and yawning. Considerably more subjects yawned while watching a thirty-minute test pattern than while watching thirty minutes of rock videos (although he didn't poll the subjects to find out which viewing experience was more bearable— we wouldn't yawn while watching and listening to thirty minutes of fingernails dragged across a blackboard, either). Did the subjects yawn for psychological reasons (they were bored) or for physiological reasons (boredom made them sleepy)?

When Provine asked his students to fill out diaries recording their every yawn, certain patterns were clear. Yawning was most frequent the hour before sleep and especially the hour after waking. And there was an unmistakable link between yawning and stretching. People usually yawn when stretching, although most people don't stretch every time they yawn.

Yawning is found throughout the animal kingdom. Birds yawn. Primates yawn. And, when they're not sleeping, fish yawn. Even human fetuses have been observed yawning as early as eleven weeks after conception. The child psychologist Piaget noted that children seemed susceptible to yawning contagion by the age of two. It was clear to Provine that yawning was an example of "stereotyped action pattern," in which an activity once started runs out in a predictable pattern. But what's the purpose of this activity?

Although Provine is far from committing himself to an answer of why we yawn, he speculates that yawning and stretching may have been part of the same reflex at one point (one could think of yawning as a stretch of the face). Bolstering this theory is the fact that the same drugs that induce yawning also induce stretching.

The ubiquity of yawning epidemics was obvious to all the people who sent in this Imponderable. Provine told *Imponderables*, "Virtually anything having to do with a yawn can trigger a yawn," and he has compiled data to back up the contention:

- 55% of subjects viewing a five-minute series of thirty yawns yawned within five minutes of the first videotaped yawn, compared to the 21% yawn rate of those who watched a five-minute tape of a man smiling thirty times.

- Blind people yawn more frequently when listening to an audio-tape of yawns.
- People who read about yawning start yawning. People who even think about yawning start yawning. Heck, the writer of this sentence is yawning as this sentence is being written.

If we are so sensitive to these cues, Provine concludes that there must be some reason for our built-in neurological yawn detectors. He concludes that yawning is not only a stereotyped action pattern in itself, but also a "releasing stimulus" that triggers *another* consistently patterned activity (i.e., another yawn) in other individuals. Yawns have the power to synchronize some of the physiological functions of a group, to alter the blood pressure and heart rate (which can rise 30% during a yawn).

Earlier in our evolution, the yawn might have been the paralinguistic signal for members of a clan to prepare for sleep. Provine cites a passage in I. Eibl-Eibesfeldt's *Ethology,* in which a European visitor to the Bakairi of Central Brazil quickly noted how yawns were accepted behavior:

> If they seemed to have had enough of all the talk, they began to yawn unabashedly and without placing their hands before their mouths. That the pleasant reflex was contagious could not be denied. One after the other got up and left until I remained . . .

Yet, Provine is not willing to rule out our evolutionary theory either. Perhaps at one time, the baring of teeth sometimes apparent in yawning could have been an aggressive act. Or more likely, combined with stretching, it could have prepared a group for the rigors of work or battle. When bored or sleepy, a good yawn might have revivified ancient cavemen or warriors.

So even if Dr. Provine can't yet give us a definitive answer to why yawning is contagious, it's nice to know that someone out there is in the trenches working full-time to stamp out Frustability. If Dr. Provine finds out any more about why yawning is contagious, we promise to let you know in the next volume of Imponderables.

Submitted by Mrs. Elaine Murray of Los Gatos, California. Thanks also to Esther Perry of Clarks Summit, Pennsylvania; Julie Zumba of Ocala, Florida; Jo Ellen Flynn of Canyon Country, California; Hugo Kahn of New York, New York; Steve Fjeldsted of Huntington Beach, California; Frank B. De Sande of Anaheim, California; Mark Hallen of Irvington, New York; Raymond and Patricia Gardner of Morton Grove, Illinois; Jim White of Cincinnati, Ohio; Renee Nank of Beachwood, Ohio; and many others.

A free book goes to Christine Dukes of Scottsdale, Arizona, for being the first to direct us to Dr. Provine.

FRUSTABLE 5: *Why Do We Give Apples to Teachers?*

This Frustable has remained remarkably resistant to reasoned replies. Although few readers could supply hard evidence to back their claims, a lot of people sure seemed to think they knew the answer to this one.

Two theories predominated. The most popular answer was the Biblical explanation. In Genesis, the forbidden fruit comes from the tree of knowledge. Although the forbidden fruit is never specified, the apple has over time been given that distinction. As Lou Ann M. Gotch of Canton, Ohio, puts it:

> the apple has come to signify knowledge. Perhaps by giving an apple to the teacher our children are admitting that they're little devils. Or perhaps they are intimating that the teachers could use a little more knowledge.

The second camp traces the custom to early rural America, when teachers were given free room and board but little pay. Students and their families traditionally brought something to

DAVID FELDMAN

contribute to the school and/or the teacher, be it wood for a fire or fruit for consumption.

But why an apple? As Georgette Mattel of Lindenhurst, New York, points out, apples were cheap and plentiful. Donald E. Saewert adds that the apple is the only fruit that can be stored for long periods of time without canning. In the winter, it might have been the only fresh fruit that was available in many areas. And Ann Calhoun of Los Osos, California, closes this argument with an impressive volley: "Sure would beat dragging a twelve-foot stalk of corn to school."

Both of these arguments are plausible but certainly not proven. Two readers sent us evidence of other possible solutions to this Frustable.

Ann Calhoun mentions that perhaps the apple-teacher connection was made up by an illustrator in one of the nineteenth-century illustrated magazines ("Every illustration I've seen . . . includes a very pretty young teacher, a blushing hayseed boy of nine, and a classroom of giggling sniggerers.")

Calhoun speculates that the boy gives the teacher an apple not as a symbol of knowledge but as a symbol of beauty. For according to ancient Greek legend, the highly prized golden apples that grew in the Garden of the Hesperus were awarded for beauty at the Judgment of Paris. Calhoun continues:

> In every Rockwellian illustration of this theme . . . the teacher is young and beautiful. Since older generations were heavily schooled in the classics (as ours is not) the teacher and kids would get the reference immediately. And yes, there are satiric variations with the teacher depicted as old, fat, gray, ugly, and scowling. The reason for her sour expression is that *she* also knows the original reference, knows the satiric content of the gesture and is about to send that little hypocritical, mendacious, miscreant presenter out behind the barn for a thrashing he so richly deserves. . . . I truly can't imagine a spindly young cleric presenting his medieval monkish tutor with an apple for beauty. Such impertinence would only earn him a thrashing outside the castle walls. So when and where did all this apple presenting and polishing start?

Good question. We've heard from only one person who dares to speculate on this. We received a fascinating letter from Henry C. Hafliger, member of the Board of Trustees of the San Jacinto Unified School District. He traces the custom back to Switzerland and cites the book *Bauerspiegel*, written by Jeremiah Gotthelf in the early nineteenth century. Gotthelf was a disciple of Johann Pestalozzi, the Swiss educator and reformer who had a tremendous influence on American education in the nineteenth century.

In effect, *Bauerspiegel* chronicles a Swiss equivalent to our rural explanation for the custom. Hafliger summarizes:

> when education was first offered to all classes of children in Switzerland, salaries of teachers were subsistence at best. Parents would supplement teachers' salaries with food, and one of the easiest foods to bring to school was the apple. Farmers would keep apples in their cellars all year long because in those days apples were not considered a dessert but a staple. Gotthelf writes that children soon learned that the child who brought the apple received the least amount of swats with the pointer or switch that the teacher always carried. The children from the poor families, some of whom did not have enough food themselves, were at a distinct disadvantage.

Hafliger's theory, then, was that the fruit was originally given on the premise that "an apple a day keeps the switch away."

Submitted by Malinda Fillingion of Savannah, Georgia. Free books go to Ann Calhoun of Los Osos, California; and Henry C. Hafliger of San Jacinto, California.

DAVID FELDMAN

FRUSTABLE 6: *Why Does Looking up at the Sun Cause Many People to Sneeze?*

Of all the Frustables, we came closest to getting a definitive solution to number six. Most of the people who responded were sun-sneezers themselves, and some said that just looking up at a bright light or even at the reflection from a car bumper was enough to set off the achoo mechanism.

The more than one hundred letters we received on this topic almost all carried some variation of the same theme. After a little digging, we found out that the most popular explanation was far from the only possible one.

We do know this much: Somewhere between 25 to 33% of the population is afflicted with "photic sneeze reflex." It is almost certainly a hereditary condition. Reader Margy D. Miller of DeKalb, Illinois, reports that she and all four of her children all sneeze when they first step out of doors into the bright sun.

The most accepted explanation for the photic sneeze reflex is that light signals that should irritate the optic nerves somehow trigger receptors that play a part in the sneezing process. The neural tracts for the olfactory and optic sensory organs lie adjacent to each other and have close (but not identical) insertion points in the brain.

When some people with this particular genetic predisposition encounter a bright light for the first time, the pupils do not contract as rapidly as they should, and the eyes are irritated. Somehow—neurologists we spoke to could not specify *how*—the olfactory and neural tracts cross-circuit.

The result: The nerves fool the brain into thinking that there is a foreign irritant in the nasal mucosa. The brain does what comes naturally; it tries to rid the sinuses of the phantom dust or pollen. The brain sends out a sneeze reflex.

Case closed?

Not quite. Reader John W. Lawrence, M.D., who specializes in internal medicine and rheumatology, gave us a variation of the above. He concurs that the original cause of the sneeze is

WHEN DO FISH SLEEP?   221

eye irritation, but believes that the tears caused by the irritation actually trigger the sneeze:

> Many people develop eye sensitivity to light. This sensitivity results in a lacrimal outburst (making of excess tears) in response to the irritation. The excess tears then run off down the lacrimal duct, which is present for this purpose. The lacrimal duct empties into the back of the nasopharynx. A drippage of liquid into the back of the nose triggers the sneeze.
>
> Only when tearing exceeds the lacrimal duct's capacity to carry runoff do tears overflow and "run down the cheeks."

William J. Dromgoole, a reader from Somerdale, New Jersey, sent us a newspaper clip indicating that scientists at Scripps Clinic and Research Foundation in La Jolla, California, have found evidence that certain allergy treatments for runny noses can cause photic sneeze reflex. Simply switching medications has rid some people of this mildly vexing problem.

Several readers have written to ask why they always seem to sneeze a particular number of times (sometimes twice, but usually three times). ENT specialists we talked to pooh-poohed it. Anyone have a theory to explain this phenomenon?

Submitted by Rick Stamm of Redmond, Washington. Thanks also to William Debuvitz of Bernardsville, New Jersey; and Lisa Madsden of Minneapolis, Minnesota.

A free book goes to James Miron, R.N., of Republic, Michigan.

　　　　　　　　　　　DAVID FELDMAN

FRUSTABLE 7: *Why Does the First Puff of a Cigarette Smell Better than Subsequent Ones?*

As we indicated in *Why Do Clocks Run Clockwise?*, the research departments of the major cigarette companies couldn't (or wouldn't) answer this Imponderable. Similarly, the Tobacco Institute and the Council For Tobacco Research—U.S.A., Inc., claimed that although much research has been conducted on the sensory awareness of cigarette smoke, this phenomenon was neither universal nor verifiable.

Luckily, *Imponderables* readers aren't as reticent as the professionals in the field. We don't have a definitive answer to this Frustable, but readers supplied us with three plausible explanations.

The Physiological Theory. Richard H. Hawkins, D.D.S., president of Medical Innovators of North America, argues that the first puff of a cigarette smells best because the olfactory nerve endings within the nasal cavity are able to interpret the smell sensation only after a rest period. "With repeated puffs, the olfaction perception goes to zero." This argument might explain why a smoker derives increasingly less satisfaction from subsequent puffs, but doesn't explain why nonsmokers, who might find cigarette smoke irritating and obnoxious, find the aroma of the first puff pleasant.

Reader Albert Wellman of Santa Rosa, California, speculates that the difference between first puffs and subsequent ones is the physical process of burning the tobacco leaves. "I suspect that once the major portion of the chemical responsible for the 'good smell' of cigarette smoke has been vaporized by the first puff of smoke, there is not enough left in the tobacco to provide a comparable olfactory experience from the remainder of the cigarette."

Wellman also hypothesizes that perhaps the olfactory nerves are temporarily blocked by some other active biochemical agent in the smoke. This theory is bolstered somewhat by research that indicates that although olfactory organs are easily fatigued, the fatigue is limited to one particular flavor. Usually, the nose will respond to a new or different smell, and there are 685 different chemical compounds found in leaf tobacco smoke.

Most of the (little) hard research we have been able to find on the sensory response to cigarette smoke doesn't corroborate these physiological explanations. Dr. William S. Cain, of the departments of Epidemiology and Public Health and Psychology at Yale University, argues that smokers don't really "taste" cigarettes in the conventional sense. The four tastes—sweet, sour, salty, and bitter—don't play much of a role in cigarette enjoyment; of the four, only the bitter is perceived by the smoker.

But Cain argues that the sense of smell is not very important either, and in the last words of the following, hints at the problem posed in this Frustable:

> it matters little for smoking enjoyment whether the smoke is exhaled through the nose or through the mouth. Smell may play a role at the moment the smoker lights up, but adaptation rapidly blunts olfactory impact.

The Tobacco as Filter Theory. Reader Jack Perkins of San Francisco, California, writes:

> As a long-time heavy smoker, I can tell you that the first puff not only smells better, it's milder. The reason for this is that the tobacco acts as a filter catching tars, nicotine, and chemicals. The further down you smoke, the greater the build-up of these substances, resulting in harsher smoke.

Rev. David C. Scott, of Bethany Presbyterian Church in Rochester, New York, agrees, adding, "The first puff has the advantage of being filtered both by the longest filter and cleanest filter. ... Each subsequent puff both shortens the filter and dirties even more what remains. Andrew F. Garruto of Kinnelon, New Jersey, compares smoking the stub of a cigarette to making a pot of coffee through used grains.

All of these arguments explain why the purity of taste and smell deteriorate as a cigarette has been smoked. But none explains to the nonsmoker why the first puff smells fine but then deteriorates immediately.

The Burning Wood, Sulfur, and Butane Theory. Even if we can't confirm any of these theories for sure, we like this modest explanation the best. Perhaps the reason why the first puff smells better is that the aroma of the lighting agent, not the tobacco, is what we are responding to. We received this letter from Allison Rosenthal, of Rancho Palos Verdes, California:

> People have always loved the smell of burning wood. By burning tree branches, pine needles, and pine cones, many not only warm their houses but improve the smell therein. If you have ever gone for a walk in Mammoth [California] in the winter, you would surely be familiar with this wonderful scent. A burning match smells much the same, maybe even a little better. Not only do you have a form of wood on a match but also sulfur, which is very pleasing when mixed with wood smoke. If you use a large 'Diamond' wood match and pull on the cigarette hard enough when lighting it, you can actually taste the sulfur and wood mixture. Even though it doesn't taste so good, it *does* smell nice.

Although Allison Rosenthal hasn't noticed that the first puff of a lighter-lit cigarette smells better, several other readers, including Judith R. Brannon of Santa Clara, California, feel that the smell of butane is the hero. As connoisseurs of gas station fumes, we would agree.

The match/lighter argument is the only theory that explains how an odor perceived as pleasant by smoker and nonsmoker alike can suddenly turn unpleasant, at least for the nonsmoker. If the hard research in sensory reactions to cigarette smoke can be believed, what a smoker perceives as a response to the taste of the flavor of a cigarette is actually a camouflage, masking a chemical response to the relief from nicotine deprivation.

A free book goes to Allison Rosenthal of Rancho Palos Verdes, California.

FRUSTABLE 8: *Why Do Women in the United States Shave Their Armpits?*

The recorded history of armpit shaving is a spotty one indeed. The earliest reference we have found was that the ancient Babylonians, more than one thousand years before the birth of Christ, developed depilatories to remove unwanted body hair.

Julius Caesar reported that the early Britons "had long flowing hair and shaved every part of their bodies except the head and upper lip," but this quotation may refer only to men. We do know that barbers removed superfluous hair from the eyebrows, nostrils, arms, and legs from male customers around this time.

The first direct reference to the specific topic at hand is contained in Ovid's *Art of Love*, written just before the birth of

DAVID FELDMAN

Christ: "Should I warn you to keep the rank goat out of your armpits? Warn you to keep your legs free of coarse bristling hair?"

In Chaucer's day (the fourteenth century), the mere sight of any hair was considered erotic. Women were required to wear head coverings; caps were worn indoors and out by women of all ages.

These ancient antecedents predict our current duality about body hair on women. On the one hand, underarm hair is considered unsightly and unhygienic, and yet on the other, sexy and natural.

None of the many razor companies or cosmetic historians we contacted could pinpoint when women first started shaving their armpits. The earliest reports concerned prostitutes during the gold rush days in California. Terri Tongco, among other readers, posited the theory that prostitutes shaved their underarms to prove they had no body lice, which were rampant in the old West.

Many older readers were able to pinpoint when their mothers and grandmothers started shaving their armpits. Not-so-old historian C.F. "Charley" Eckhardt of Seguin, Texas, is the only person we have found who has actually studied this Frustable:

My paternal grandmother, born in 1873, and my maternal grandmother, born in 1882, did not shave their armpits. My wife's maternal grandmother (1898), my mother (1914), and my mother-in-law (1921) all did or do.

Eadweard Muybridge's photographic studies of the nude human figure in motion and Hillaire Belloc's photographs of New Orleans prostitutes, all taken before or immediately after the turn of the century, show hairy armpits, as do nude photos of prostitutes known to have been taken in El Paso, Texas, prior to 1915. In addition, still photographs taken from pornographic motion pictures known to have been made prior to 1915 show the women with unshaven armpits, as do surviving pornographic photographs of the "French postcard" variety which are documented as having been made in the United States prior to 1915.

Theatrical motion pictures released about and after 1915, including *Cleopatra* (starring Theda Bara), the biblical sequences from D.W. Griffith's *Intolerance,* and several others, show shaven armpits. Something, then, happened about 1915 that would cause not merely stars but impressionable teenagers (as my wife's grandmother was) but not necessarily older family women (like my grandmothers) to start shaving their armpits.

So what caused these women to start shaving their armpits around 1915? Many readers, including Charley Eckhardt, give the "credit" to Mack Sennett:

> The first moviemaker to show the feminine armpit extensively in non-pornographic films was Mack Sennett, in his Bathing Beauty shorts . . . Sennett's Bathing Beauties had shaven armpits, and they are the first direct evidence we have of the armpit-shaving phenomenon. Whether or not Mack actually said 'That looks like hell—have 'em shave' is a moot point, though the statement is completely in character with what we know about Sennett.

We do know that flappers of the Roaring Twenties adopted the sleeveless clothing that seemed so daring in the Sennett shorts.

We heard from several women who were more concerned about why the custom persists rather than how and when it started. Typical was this letter from Kathy Johnson of Madison, Wisconsin:

> I am one of the apparently few U.S. women who has never shaved her armpits or legs. It never made logical sense to me, so why do it? I've heard the argument that shaving those regions is more sanitary. Then why, I volley back, don't men shave their armpits? Why, in fact, doesn't everyone shave their heads if lack of hair is so sanitary? Stunned silence . . .

Several psychologists and feminists have speculated that men like the shaven look because it makes women look prepubescent —young, innocent, and unthreatening. Diana Grunig Catalan of Rangely, Colorado, who subscribes to the prepubescent theory, speculates that "American women, unlike their European counterparts, were not supposed to do anything with all those men they attracted with their revealing clothing. A childlike, helpless look can be a protection as well as an attractant."

DAVID FELDMAN

In defense of men, it has been our experience that many women have visceral reactions to the presence or lack of body hair in men. Why does the same woman who likes hair on the front of the torso (the chest) not like it on the back? Why is hair on the arms compulsory but excess hair on the hands considered repugnant? Are women, as well as men, afraid to face the animal part of our nature? Hairy questions, indeed.

Submitted by Venia Stanley of Albuquerque, New Mexico.

A free book goes to C.F. "Charley" Eckhardt of Seguin, Texas.

FRUSTABLE 9: *Why Don't You Ever See Really Tall Old People?*

This Imponderable-turned-Frustable was submitted by Tom Rugg, who stands six foot six inches and understandably has a vested interest in the answer.

Many readers sent us lists of reasons why people get shorter as they get older. Some of the reasons include gravity; the degeneration, rigidification, and compression of the vertebral column as we get older; osteoporosis; curvature of the spine. All of these phenomena explain why we might lose two or three inches over a lifespan, but don't explain why we haven't seen the six-foot-nine person who has "shrunk" to six foot six.

Several people wrote to say that improved nutrition has made our population taller than it used to be. Presumably, our generation will grow old and "really tall" with a lifetime of Twinkies and Diet Coke in our systems. Yes, we have grown taller but on average little more than a half inch in the last twenty-five years and fewer than two inches since the beginning of the century.

Dr. Alice M. Mascette of Tacoma, Washington, and Cindy West of Towson, Maryland, mentioned that a portion of our really tall population is afflicted with Marfan's syndrome, a genetic affliction of the connective tissue of the body. Sufferers of Marfan's syndrome have abnormally large hands and feet and a subpar heart. Many die of a ruptured aorta after an aneurism.

So far, these Marfan's syndrome sufferers—only a small fraction of all very tall people—are the only identifiable group of tall people who have been proven to have a short lifespan. But it is not at all clear that the tallness per se is what causes their deaths.

The way to unlock this Frustable is by asking: Do very tall people have shorter lifespans than other people? Surprisingly, there is no scientific data to support the proposition. We heard from more than fifteen doctors, health agencies, and insurance companies, and none of them study mortality based on height alone. Metropolitan Life conducts countless studies on the relationship between height-weight ratios and longevity, but doesn't feel that there is any reason to believe that tall people have a higher morbidity rate than the population as a whole.

In fact, the only quasi-scientific study we've seen (sent to us by reader David Jordan) that claims that very tall people live shorter lives was conducted by an aerospace engineer, Thomas T. Samaras. He tracked the lifespans of three thousand professional baseball players and found that the tallest players (six foot six or taller) lived, on average, to only the age of fifty-two. On the other hand, the shortest group (under five foot four) lived more than sixty-six years on average.

All of the medical and insurance experts we spoke to doubted the validity of Samaras' results, as well as his reasoning. Samaras speculated that the heart of a tall person must work overtime to pump blood a longer distance than a short person. Johns Hopkins University heart specialist Dr. Solbert Perlmutt disagreed with this argument and added, "Besides, you don't see mice living long. But you see elephants doing quite well."

DAVID FELDMAN

And evidently some old people only slightly shorter than elephants are doing pretty well, too, though the scarcity of the really tall old person is evidenced by the fact that of the hundreds of thousands of people who read *Why Do Clocks Run Clockwise?* only one person stepped up to the plate and offered himself as a specimen. Robert Purdin of Tinton Falls, New Jersey, is sixty-five (is that old?) and six foot five (is that really tall these days?).

Dr. Emil S. Dickstein of Youngstown, Ohio, says that he sees many tall old people, as does Gwen Sells, a member of Tall Clubs International. Reader George Flower, who once encountered a six foot seven man in his seventies, reminds us that Jimmy Stewart, if not "really" tall, is pretty tall.

But our favorite sighting was sent in by Andy Stone of Denver, Colorado, who told us about Randy "Sully" Sullivan, who weighs trucks at the Port of Entry in Cortez, Colorado:

> Sully is six foot ten inches. I've never asked his age but his hair is white, his posture stooped (that's right, stooped), and I estimate he's about seventy.

So, Tom Rugg, there's hope for you yet.

Submitted by Tom Rugg of Sherman Oaks, California.
Thanks also to Joanna Parker of Miami, Florida.

A free book goes to David Jordan of Greenville, Mississippi.

FRUSTABLE 10: *Why Do Only Older Men Seem to Have Hairy Ears?*

How appropriate that we saved the most frustrating Frustable for last. In *Why Do Clocks Run Clockwise?*, we mentioned that we consulted endocrinologists who professed ignorance on the subject. "If this condition is only found in males, why don't you speak to geneticists," they chimed in unison.

So we talked to geneticists. Guess what they said.

"Why don't you talk to endocrinologists? They'd know about this stuff."

So we put this in as our last Frustable and waited for the mail to roll in. It did.

Most of the mail had much the same answer as this one, from Bryan, Texas:

> I am Stacey Lero, a seventh grader at Anson Jones School. . . . We are studying genetics in science. On the Y chromosome there is a gene for hairy ears. It matures throughout your lifetime and as men reach the late forties or early fifties (sometimes earlier or later), it has matured enough to be expressed and the hair begins to grow. I hope I've answered your question.

What??? Heads of genetic departments at prestigious universities can't answer this Imponderable and Stacey Lero is studying it in a seventh grade science class? What's going on here?

Then several other readers, including Richard Landesman, associate professor of zoology at the University of Vermont, and R. Alan Mounier of Vineland, New Jersey, sent me clips from genetics textbooks that confirmed Stacey Lero's letter. One text said that hypertrichosis (excessive hair) of the ear is passed directly from father to son.

Feeling humbled by the knowledge of our readers, we consulted some more geneticists. They replied that the Y-chromosome theory had been largely discounted—no hard research supports this belief. "Why don't you talk to an endocrinologist," said one soothingly.

Are we the only ones who feel a little queasy about medical textbooks printing untrue facts? Or are scientists and doctors not believing what is in medical textbooks?

Peter H. Lewis, a reporter at the *New York Times*—the paper of record, for darn's sake—called us excitedly to say that they had run an article in 1985 about hairy ears being signs of

DAVID FELDMAN

susceptibility to heart attacks. In 1984, two doctors in Mineola, Long Island, reported to the *New England Journal of Medicine* that there was a "significant statistical link" between men (but not women) who had hair in their ear canal and people they had treated for coronary artery disease. The doctors did not overplay the significance of this finding. In fact, the hubbub their findings released prompted Dr. Richard F. Wagner and Dr. Karen Dineen to issue a poetic disclaimer:

> If on the ear there is a crease
> Do not assume that life will cease.
> If hair is noted in the ear,
> Do not assume that death is near.
> So, if when walking down the street
> An ear with hair and crease you meet,
> Don't give the gent a dreadful fright—
> Don't hint infarction is in sight.

Needless to say, the medical authorities we consulted would neither affirm nor deny the viability of the androgen theory.

We give up. Some Frustables are too frustrating even for us, and we're masochists.

We figure that Stacey Lero will be going to high school soon. She's obviously very bright and will probably become a science major in high school. She will then enter college, where she will become a double endocrinology/genetics major. She'll choose between MIT and Cal Tech for her graduate work. In the year 2011, Stacey will win the Nobel Prize for answering this Frustable. The world will be a better place. And it will all be due to that seventh-grade science teacher in Bryan, Texas. Well, and maybe a little to the inspiration provided by that free copy of *When Do Fish Sleep?*

A free book goes to Stacey Lero of Bryan, Texas.

We have received more than two thousand letters since the publication of *Why Do Clocks Run Clockwise?*. Most of them posed Imponderables or tried to answer Frustables. But some corrected or added information to our answers or contained priceless comments about the topics in our first two volumes of *Imponderables*. Here are some of our favorites.

On the Relative Merits of Round vs. Flat Toothpicks

We commented that even the manufacturers of flat toothpicks couldn't provide any reason why they were superior to round toothpicks, except for their lower price. Who would have thought that this topic could rouse emotions? Some letters were thoughtful, others passionate.

> Flat toothpicks have uses round ones don't have, such as smearing small amounts of various kinds of goo onto surfaces (epoxy cement is one) and in being able to enter crevices closed to round toothpicks. Flat picks, with their greater surface, are in my experience superior to round ones for testing doneness of cakes, custards and so on. . . .
>
> MAX HERZOG
> *Augusta, Georgia*

> When I read your slam at flat toothpicks in *Imponderables*, I thought, "Gee, I hope I can find this guy's address so I can straighten him out." And lo and behold, there it was. You even invited comment. You must be a brave man. If you can't stand the thought of reading a defense of flat toothpicks, skip this part.
>
> I *hate* round toothpicks, The damn things are too fat and too close to their little pointy ends, which means that they won't go between my teeth at the gum line far enough to push out the bits of whatever gets stuck in there. Flat toothpicks will. So what if I have to throw a few away when they break before I can accomplish much; they're cheap, as you pointed out.

Furthermore, being pointed on both ends, round toothpicks are worthless for polishing the front surfaces of my teeth, unless I chew them down a bit first. I think the makers of flat toothpicks should square off those big ends, but at least those ends aren't pointy as on those round jobbers, and I can do some good with them.

Whenever a restaurant has only round toothpicks, I take out my pocket knife and whittle 'em down so they'll go through the gaps. I haven't worked up the nerve yet to do this in front of the maître d', but one of these days I will, scattering toothpick slivers on the carpet, to make sure the message is absorbed.

Alan M. Courtright
Seattle, Washington

On Why Countdown Leaders on Films Don't Count Down to One

Your information is correct until you reported that the number one is the start of the picture. Although there isn't a one on Academy Leader, the picture actually starts on what would be zero. The forty-seven frames of black film that follow the single frame bearing a "2" are for the projectionist to open the dowser and allow light through the projector. A quick "beep" is usually heard along with the number two, indicating that the sound is in sync with the picture.

In theaters that alternate between two projectors, there is a mark that appears in the upper right-hand corner of the picture, which tells the projectionist to start the other projector up to speed, and then a second mark, which is when the projectionist actually should change over to the new reel. This countdown leader allows a precise amount of time for the projector to get up to speed, so that when the changeover occurs the viewer will not have missed any of the movie.

Brian M. Demkowicz
Chief Projectionist, IMAX Theater
Baltimore, Maryland

DAVID FELDMAN

On Why American Elections Are Held on Tuesday

Election day is not the second Tuesday in November but is the first Tuesday after the first Monday in November.

STEVEN J. RIZZO
Islip, New York

On Why Balls Are on Top of Flagpoles

I was always taught that the answer was longevity of the pole. In the days when flagpoles were wooden, the end grain of the wood was exposed if not capped by a ball or other type of finial. End grain absorbs dampness more readily than any other part of the wood. . . .

GAVIN DUNCAN
Tabb, Virginia

Several veterans, including retired Army Sergeant Robert E. Krotzer of Hephzibah, Georgia, wrote to say that they were taught that the purpose of the ball was to keep the flag from being caught on the pole when the wind blew the flag upward. The flag experts we spoke to admit that this is the reasoning the Army provides, but insist that even a sphere doesn't stop a flag from getting stuck on top of a pole.

On Why the Sound of Running Water Changes When Hot Water Is Turned On

I do not deny the validity of the causes you discussed, however the *pitch* of that sound at a given rate of flow depends on the density of the water. Hot water is substantially less dense than cold water. . . . The fundamental fact of physics is that different density fluids have different natural frequencies of vibration while flowing through a given orifice.

Just turn on any hot-water faucet that has been off long enough for the water content to get cold some distance down the pipe. Then stand back and listen. You will clearly hear a change in pitch as the hot water arrives. The change is sudden and cannot be explained by any adaptive change of the pipes. It is the direct result of the change in the natural frequency of the water itself. The noisier the flow the more noticeable the change.

STEFAN HABSBURG
Farmington Hills, Michigan

On Why We Aren't Most Comfortable in 98.6° F Temperature

In *Why Do Clocks Run Clockwise?*, you wrote we would feel most comfortable when it is 98.6° F in the ambient air—if we were nudists.

Not *exactly* so. Human beings use up caloric energy, derived from food, to make motions with our muscular bodies. This process yields a certain amount of excess energy in the form of heat. Our bodies radiate this excess heat into the ambient air. When severely overheated, our bodies hasten the action by evaporating sweat. But we must have a means to *keep* our body temperature at 98.6° F or we die of heat prostration.

If we were all nudists and the ambient air *everywhere* was 98.6° F, we'd feel discomfort the moment we began to move. Lacking a temperature differential in the ambient air, our bodies could no longer radiate heat easily. First we'd sweat, and then we'd all die.

The only hope to remain alive would be to remain as motionless as possible for as long as possible, but sooner or later the excess heat from involuntary motions (like the heart and lung muscles) would build up.

So a 98.6° F temperature wouldn't be "comfortable" very long. . . . We need a slightly lower temperature in the air sooner or later.

DON SAYENGA
Bethlehem, Pennsylvania

DAVID FELDMAN

On Bird Droppings

You can't get away with anything with Imponderables *readers. We simplified a little by calling the white stuff surrounding the black dot in bird droppings "urine." One reader noticed.*

Mammals and amphibians get rid of nitrogenous waste in the form of urea dissolved in water. This is the material we commonly call urine. Birds and reptiles cannot accomplish this. They get rid of their nitrogenous wastes in a white semisolid form called uric acid. This is the white material in the birds' droppings.

There are two reasons why birds and reptiles use uric acid for waste disposal. One is because it is a water conservation technique. The other reason is perhaps more important. Bird and reptile embryos develop inside a hard shell. If they were to produce water-soluble urea while developing, it would end up poisoning the embryo before it could fully develop and hatch.

This leads us to the answer of another interesting question, "What is that 'gooky' stuff inside the shell after a baby bird hatches?" It is the remains of what is called the "allantois," the garbage can where nonsoluble uric acid is deposited while the embryo is developing.

SANDY JONES
Manassas, Virginia

On the Purpose of the Half Moons on Fingernails

Although our explanation—lunula are trapped air and serve no biological purpose—was correct, one reader did find a way to use them:

When preparing a patient prior to surgery requiring a full anesthetic, I was told to remove all nail polish prior to admission. When I asked why, I was told that recovery room personnel can monitor blood pressure by observing changes in the color of half moons.

MICHEALE WILLIAMS
Portland, Oregon

WHEN DO FISH SLEEP?

On the Mysterious Fruit Flavors Contained in Juicy Fruit Gum

When I was in college, I made the synthetic flavors of oil of pine-apple (ethyl butyrate) and oil of banana (amyl acetate). I found when mixed in precisely a certain ratio, I got the distinct aroma of Juicy Fruit. . . .

Incidentally, if one wishes to synthesize ethyl butyrate, be prepared. Butyric acid is one stinking, sickening smelling acid. But once mixed with ethyl alcohol and concentrated sulfuric acid, the ethyl butyrate emerges with a sweet pineapple aroma.

HAROLD E. BLAKE
Tampa, Florida

On the Purpose of Pubic and Underarm Hair

Most of the experts we contacted speculated that this body hair served as a sexual attractant. But in a letter to Human Evolution, *one reader dissented. We reprint part of the letter with his permission:*

Pubic and axillary hair have been assumed to be biologically non-functional and therefore relegated to a role of mere sex attractants or to signal sexual maturity. Yet if one examines the action of axillary and pubic hair it can be seen that these patches serve as a kind of lubricant for arm and leg movements repectively and must have been retained in that capacity or evolved separately when other body hair was lost. One can easily observe the friction-reducing function of axillary hair by shaving under one arm and noting the added friction of the shaved arm. The fact that pubic hair extends up the abdomen beyond the point where it facilitates leg movement may mean that body hair was lost while our fore-bears were still walking in a crouch or on knuckles; for it comes into function, particularly the lateral portions, in that position. It would serve well for a semicrouched or sometimes-crouched proto-hominid that had lost most of its body hair. As our ancestral mothers began losing their body hair, fatty breasts and pubic and axillary hair could have all evolved simultaneously. The hair patches were selected for the purely biological function of reducing friction whereas general loss of body hair gave rise to the

DAVID FELDMAN

necessity of fatty breasts for providing the crucial psychological role of softness, comfort, and security for the infant.

NOEL W. SMITH
State University of New York
Plattsburgh, New York

On Why Ranchers Hang Old Boots Upside-down on Fence Posts

The longest chapter in Why Do Clocks Run Clockwise? *was a futile attempt to answer this Imponderable. We confirmed that Nebraska was the epicenter of boot-hanging activity. We even found the son of the man reputed to have started the practice. But even he didn't know why his father hung the boots. Some readers had their own ideas.*

Marla Bouton, of Kearney, Nebraska, sent us an article by Roger L. Welsch, professor of English and anthropology at the University of Nebraska–Lincoln, published in the October 30, 1983, Sunday World-Herald Magazine *of the Midlands. Along with repeating all of the theories we advanced, Welsch recounted many other stories he was told by boot-hangers, including the number of boots indicated the number of sons in the family; the toes point toward the nearest graveyard; the toes point toward the main house in case someone was lost in a snowstorm; and the boots are a token of good luck.*

Welsch concludes that although there may not be one single answer, hanging boots is probably some form of territorial marker. He notes that boot hanging is most prevalent in arid flatlands.

> In a geography like this, long arrays of boots are striking, even stunning, and that is precisely their purpose. They are markers. They announce that someone lives here in this moonscape, that there are inhabitants, no matter how "deserted"—a perfect word, "deserted!"—things appear to be. . . .

Several readers insisted there was a more practical explanation for the custom.

The ranchers may be trying to stop the absorption of water. . . . In Alabama, a lot of farmers turn empty cans onto the tops of fence posts for this reason, or they will nail the tops that were taken from cans, onto the tops of the posts. This keeps the posts from absorbing large amounts of water when it rains. Wooden posts absorb quite a bit of water through the tops. Putting boots on the posts might prevent the wood from rotting prematurely.

C.A. "JUNIOR" WEAVER
Millbrook, Alabama

In West Virginia, some of the older farmers, including my grandparents, used to put on tin cans, old pieces of tires, roof shingles, or something else that would cover the top and hang down the sides of the fence posts.

This practice was done mostly to fence posts that still had bark on them. The farmers felt there was no reason to put objects on posts that had the bark stripped off.

Still asking why? Believe it or not, the reason was to keep the fence posts from rotting.

The idea was to keep the rain and snow from laying on the top of the post and soaking or running behind the bark. They believed the rain or snow would run down behind the bark, become trapped and rot the wood faster than if there was no bark at all on the posts. . . .

My husband and I have fence posts in our backyard (they are over 10 years old) and the bark has been stripped off. They show no signs of rot so far. They are so hard you can't hardly drive a nail into them.

Every once in awhile when I am traveling on some of the older, less busier country roads in the state, I see a fence with something on the top of the fence posts, and I remember asking my grandfather why he was doing it. I am glad that I was curious enough to ask because I may have helped you solve an Imponderable that seemed to be driving you nuts.

ELAINE K. SOUTHERN
Clarksburg, West Virginia

DAVID FELDMAN

If most of our letters on this subject came from the South and the Midwest, we received at least one sighting considerably farther to the north.

I was very surprised to see this in your book as I thought this practice was only done in my old territory.

I was a district rep for a car manufacturer and my territory included the central east of Alberta, Canada. This included Drumheller and Trochu, two small towns on either side of the Deer River.

Drumheller is famous for being the site of one of the first large dinosaur finds in North America. It now has a large scientific museum that attracts thousands of visitors every year. Trochu is not famous for anything, although it does have a very good ice-cream stand open during the summer.

Anyway, the back road from Drumheller to Trochu is one of the most pleasant drives you can find on the prairie. . . .

After crossing the river and driving toward Trochu and the ice-cream stand, there is a rancher who has put hundreds of boots on his fencing along the road. I asked about them and was told that they were to stop the aging of the fence posts. If the tops of the posts are covered and not left exposed, they will last that much longer. And since I once had a job replacing an old fence, I can assure you that anything that can be done to make them last longer would be tried.

KEVAN TAYLOR
Niagara Falls, Ontario, Canada

Acknowledgments

The single most gratifying part of my job is receiving the thousands of letters that readers of *Imponderables* have sent me. Your ideas have supplied most of the mysteries answered in this book. Your support and encouragement have supplied the inspiration.

I have kept my promise to answer all letters that have included self-addressed stamped envelopes. I'll continue the practice, but please be patient. When deadline pressure mounts, so does my response time. I cherish your letters and pounce on each one like a child encountering a wrapped birthday present.

In Harper & Row, I have a found a publishing house that provides me with all the benefits of a family—without the in-laws. Excepting that he is taller and wears clothes better than me, Rick Kot is all I could ask for in a person or an editor. His assistant Scott Terranella is exhibiting annoying tendencies toward becoming as perfect as Rick, but Scott has been so kind to me it's hard to get mad at him.

From the top down the folks at Harper & Row have been gratuitously nice to me. The publisher, Bill Shinker, has been constantly supportive and enthusiastic. The beloved Brenda Marsh and the sales reps (sounds like a Motown act!) have, wonder of wonders, gotten my books into the stores. Roz Barrow, with skill and graciousness, made sure there were enough books to ship to the stores. Steve Magnuson has been full of great marketing ideas. Debra Elfenbein, with a sharp mind and several sharp red pencils, helped tighten and focus this manuscript. The publicity department, headed by Karen Mender, helped thrust me upon an innocent North America. Special thanks to my publicist and rock 'n' roll heartthrob, Craig Herman, and to Allison Koop, Susie Epstein, and Anne Berman. And to the trinity in Special Markets, Connie Levinson, Barbara Rittenhouse, and Mark Landau: You have a friend for life, whether you like it or not.

In *Why Do Clocks Run Clockwise?*, I complained that my agent, Jim Trupin, didn't laugh enough at my jokes. I'm happy and proud to announce that he has corrected this egregious flaw and can now lay claim to be the last Renaissance man. Jim and his wife, Elizabeth, are two of my favorite people. Speaking of favorite people, Kassie Schwan, illustrator and semiprofessional gardener, continues to produce terrific illustrations. And the late (not dead, just late) Mark Kohut has taught me more than anyone about how the book business works. Lovely Joann Carney is the only person who has ever gotten me to sit in front of a camera for more than five minutes without wiggling uncontrollably, let alone to enjoy the process of being photographed.

Over the last few years, I've had a chance to meet an underpaid, unsung but fabulous group of people—booksellers. From the president and CEO of Waldenbooks, Harry Hoffman, to the managers of mall stores, Julie Lasher and Brian Scott Rossman; from B. Dalton's manager of Merchandise Planning and Communication, Mattie Goldberg, to all the folks at the Benjamins Bookstore in the Pittsburgh Airport, the booksellers I've met have been intelligent, committed, and inordinately good company. Thanks for providing me with an on-the-job education.

My friends and family have helped me survive a difficult year. Thanks to all who have lent support: Tony Alessandrini; Michael Barson; Rajat Basu; Ruth Basu; Jeff Bayone; Jean Behrend; Brenda Berkman; Cathy Berkman; Sharon Bishop; Carri Blees; Christopher Blees; Jon Blees; Bowling Green State University's Popular Culture Department; Jerry Braithwaite; Annette Brown; Arvin Brown; Herman Brown; Joann Carney; Janice Carr; Alvin Cooperman; Marilyn Cooperman; Judith Dahlman; Paul Dahlman; Shelly de Satnick; Linda Diamond; Joyce Ebert; Steve Feinberg; Fred Feldman; Gilda Feldman; Michael Feldman; Phil Feldman; Phyllis Fineman; Kris Fister; Linda Frank; Seth Freeman; Elizabeth Frenchman; Michele Gallery; Chris Geist; Jean Geist; Bonnie Gellas; Bea Gordon; Dan Gordon; Ken Gordon; Judy Goulding; Chris Graves; Christal Henner; Marilu Henner; Melodie Henner; David Hennes; Paula Hennes; Sheila Hennes; Sophie Hennes; Steve Hofman; Uday Ivatury; Terry Johnson; Sarah Jones; Mitch Kahn; Dimi Karras; Mary Katinos; Robin Kay; Stewart Kellerman; Harvey Kleinman; Mark Kohut; Claire Labine; Randy Ladenheim-Gil; Debbie Leitner; Jared Lilienstein; David Lynch; all my friends at the Manhattan Bridge Club; Phil Martin; Jeff McQuain; Julie Mears; Phil Mears; Carol Miller; Barbara Morrow; Phil Neel; Steve Nellisen; Millie North; Milt North; Charlie Nurse; Debbie Nye; Tom O'Brien; Pat O'Conner; Joanna Parker; Jeannie Perkins; Merrill Perlman; Joan Pirkle; Larry Prussin; Joe Rawley; Rose Reiter; Brian Rose; Paul Rosenbaum; Carol Rostad; Tim Rostad; Susie Russenberger; Leslie Rugg; Tom Rugg; Gary Saunders;

Joan Sanders; Mike Sanders; Norm Sanders; Cindy Shaha; Patricia
Sheinwold; Kurtwood Smith; Susan Sherman Smith; Chris Soule;
Karen Stoddard; Kat Stranger; Anne Swanson; Ed Swanson; Mike
Szala; Josephine Teuscher; Carol Vellucci; Dan Vellucci; Hattie
Washington; Julie Waxman; Roy Welland; Dennis Whelan; Devin
Whelan; Heide Whelan; Lara Whelan; Jon White; Ann Whitney; Carol
Williams; Maggie Wittenburg; Karen Wooldridge; Maureen Wylie;
Charlotte Zdrok; Vladimir Zdrok; and Debbie Zuckerberg.

Well more than one thousand educators, institutions, experts,
foundations, corporations, and trade associations were contacted for
this book. Because we can't go to reference books to get our answers to
Imponderables, we are dependent upon the generosity of the folks
listed below. Although many other people supplied help, those listed
below gave us information that led directly to the solution of
Imponderables in this book: Sandra Abrams, Associated Services for
the Blind; Richard B. Allen, Atlantic Offshore Fishermen's
Association; Dr. Robert D. Altman, A & A Veterinary Hospital;
American Academy of Dermatology; American Hotel and Motel
Association; Carl Andrews, Hershey Foods; Richard A. Anthes,
National Center for Atmospheric Research; Gerald S. Arenberg,
National Association of Chiefs of Police; Dr. Edward C. Atwater,
American Association for the History of Medicine.

Dr. Don E. Bailey, American Association of Sheep and Goat
Practitioners; Dr. Ian Bailey, School of Optometry, University of
California, Berkeley; Jan Balkin, American Trucking Associations; Dr.
Pat A. Barelli, American Rhinologic Society; Nancy Beiman, National
Cartoonists Society; Roy Berces, Pacific Stock Exchange; Dr. William
Berman, Society for Pediatric Research; Dr. William Bischoff,
American Numismatic Society; Ed J. Blasko, Eastman Kodak
Company; Dr. Peter Boyce, American Astronomical Society; Richard
Brooks, Stouffer Hotels; Edwin L. Brown, American Culinary
Federation; Bureau of the Mint, Department of the Treasury; Dr.
Walter F. Burghardt, American Veterinary Society of Animal Behavior;
Herbert H. Buzbee, International Association of Coroners and Medical
Examiners.

John Canemaker; Gerry Carr, International Game Fish
Association; Carel Carr, Yellow Pages Publishers Association; Helen
Castle, Kellogg's; Louis Chang; Bob Cochran, Society of Paper Money
Collectors; Linden Cole, Society of Actuaries; Linda W. Coleman,
Department of the Treasury, Bureau of Engraving and Printing; Robert

L. Collette, National Fisheries Institute; Dr. James D. Conroy, College of Veterinary Medicine; Charles T. Conway, Gillette Company; Philip S. Cooke, Inflight Food Service Association; Captain K.L. Coskey, Navy Historical Foundation; Louise Cotter, National Cosmetology Association; Danny J. Crawford, Marine Corps Historical Foundation; Edward Culleton, Green Olive Trade Association.

Hubert R. Dagley II, American College of Sports Medicine; Paul N. Dane, Society of Wireless Pioneers; Neill Darmstadter, American Trucking Association; Dr. Frank Davidoff, American College of Physicians; Professor Michael De L. Landon, American Society for Legal History; Brian M. Demkowicz, Maryland Academy of Sciences; Dr. Liberato John A. DiDio, International Federation of Associations of Anatomists; James J. Donahue, Duracell Inc.; Richard H. Dowhan, GTE Products; Don R. Duer, Still Bank Collectors Club of America; Thomas Dufficy, National Association of Photographic Manufacturers; W.K. Bill Dunbar, Morse Telegraph Club.

Susan Ebaugh, Serta Inc.; Dr. William G. Eckert, INFORM; Carole L. Edwards, Mobil Oil Corporation; Peter Eisenstadt, New York Stock Exchange Archives; Kay Engelhardt, American Egg Board.

Raymond E. Falconer, Atmospheric Sciences Research Center, SUNY at Albany; Dr. Fred Feldman; Dr. Barry Fells, Epigraphic Society; Stanley Fenvessey, Fenvessey Consulting; Peter C. Fetterer, Kohler Company; Deidre Flynn, Popcorn Institute; Bruce A. Foster, Sugar Industry Technologists; Don French, Radio Shack; Lester Frey, Villamarin Guillen.

Samuel R. Gammon, American Historical Association; Dr. James Q. Gant, International Lunar Society; Bruce R. Gebhardt, North American Native Fishes Association; Chris George, Rand McNally; Gerontology Research Center, National Institute of Aging; Karen L. Glaske, United Professional Horsemen's Association; Jacqueline Greenwood, Black & Decker; Patricia A. Guy, Bay Area Information System Reference Center.

Susan Hahn, United States Tennis Association; Dr. John Hallett, Desert Research Institute; Korynne Halverson, Evans Food Group; David A. Hamilton, Professional Ski Instructors of America; Lynn Hamlin, National Syndications Inc.; Darryl Hansen, Entomological Society of America; Carl Harbaugh, International Association of Chiefs of Police; Dorcas R. Hardy, Commissioner of Social Security; John Harrington, Council for Periodical Distributors; Tamara J. Hartweg, Kraft; Connie Heatley, Direct Marketing Association; Jim Heffernan,

ACKNOWLEDGMENTS

National Football League; Richard Heistchel, Schinder Elevator Company; Jacque Hetrick, Spalding Sports Worldwide; Shari Hiller, Sherwin-Williams Company; Janet Hinshaw, Wilson Ornithological Society; Robert C. Hockett, Council for Tobacco Research; Dick Hofacker, AT&T Bell Laboratories; Greg Hoffman, Jolly Time; Beverly Holmes, Frito-Lay Inc.; Dr. Daniel Hooker, University of North Carolina at Chapel Hill, Student Health Service; Richard H. Hopper; Donald Hoscheit, Osco Drug; Mark R. Houston, California Kiwifruit Commission; Professor Barbara J. Howe, National Council on Public History; Kenneth Hudnall, National Yellow Pages Agency Association; Hyde Athletic Industries, Inc.

Dr. Peter Ihrke, American Academy of Veterinary Dermatology; International Bank Note Society; Helen Irwin, National Tennis Hall of Fame.

John Jay, Intercoiffure America; Dr. William P. Jollie, American Association of Anatomists; Larry Josefowicz, Wilson Sporting Goods Company.

Jeff Kanipe, *Astronomy*; Robert Kaufman, Metropolitan Museum of Art; Edward E. Kavanaugh; Dr. Thomas P. Kearns, American Ophthalmological Society; Michele Kelley, American Hotel and Motel Association; Dr. Anthony L. Kiorpes, University of Wisconsin, Madison, School of Veterinary Medicine; Dan Kistler, Christian Research Institute; Dr. Ben Klein; Samuel Klein, United States Postal System; Ken Klippen, United Egg Producers; Dr. Kathleen Kovacs, American Veterinary Medicine Association; Stanley Kranzer, Metropolitan Life.

Jean Lang, Fieldcrest; Keith Lattislaw, National Center for Health Statistics; John Laughton, American Standard; Mary Jane Laws, American Dairy Association; Cathy Lawton, Shulton Inc.; Dr. Beverly Leffers, Milton Helpern Institute of Forensic Medicine; Professor Alfonz Lengyel, Eastern College; Dick Levinson, H.Y. Aids Group; Peter H. Lewis, *New York Times*; Pierre Lilavois, New York City Sewer Department; Barbara Linton, National Audubon Society; Kenneth M. Liss, Liss Public Relations; John Loftus, Society of Collision Repair Specialists; Joan G. Lufrano, Foote, Cone & Belding; Lynne Luxton, American Foundation for the Blind.

William L. MacMillan III, Pencil Makers Association; Alan MacRobert, *Sky & Telescope*; Dr. M. Mackauer, Center for Pest Management; Joseph D. Madden, Drug, Chemical and Allied Trades Association; Reverend Robert L. Maddox, Americans United for

Separation of Church and State; Mail Order Association of America; William C. Mailhot, Gold Medal Flour; Michael Marchant, Ogden Allied Aviation Services; Ginny Marcin, Campbell Soup Company; Colonel Ronald G. Martin; Howard W. Mattson, Institute of Food Technologists; Dr. Robert McCarley, Sleep Research Society; James P. McCauley, International Association of Holiday Inns; Dr. Everett G. McDonough, Zotos International; William F. "Crow Chief" Meyer, Blackfeet Indian Writing Company; Mary D. Midkiff, American Horse Council; Jerry Miles, American Baseball Coaches Association; David G. Miller, National Association of Retail Druggists; Dr. Stephen Miller, American Optometric Association; Robert J. Moody, General Electric; Rita Moroney, Office of the Postmaster General; Pete Morris, C.H. Morse Stamp Company; Bill Mortimer, Life Insurance Marketing and Research Association; George Motture, Wise Foods; Meg Wehby Muething; Arthur J. Mullkoff, American Concrete Institute; Edith Munro, Corn Refiners Association; Gordon W. Murrey, Murrey International; D.C. Myntti, American Bankers Association.

Dr. David Nash, American College of Physicians; National Institute on Aging, National Institute of Health; Arnie Nelson, Yellow Spots; David Nystrom, U.S. Geological Survey.

Norman Oehlke, International Fabricare Institute; Carl Oppedahl.

Dr. Lawrence Charles Parish, History of Dermatology Society; Dianne V. Patterson, United States Postal System; William R. Paxton, Federal Railroad Administration; Peggy Pegram, Bubble Yum; Joy Perillo, AT&T Archives; Pillsbury Company; Leslye Piqueris, American Foundation for the Blind; Lawrie Pitcher Platt, Tupperware Home Parties; Bruce Pluckhahn, National Bowling Hall of Fame and Museum; Proctor-Silex; Dr. Robert Provine, University of Maryland; Roy S. Pung, Photo Marketing Association, International; Thomas L. Purvis, Institute of Early American History and Culture.

Jerry Rafats, National Agriculture Library; Dr. Salvatore Raiti, National Hormone and Pituitary Program; Monika Reed, Berol USA; Walter Reed, National Automatic Merchandising Association; Al Rickard, Snack Food Association; Bob Riemer, Gasoline and Automotive Service Dealers; R.J. Reynolds; Robert S. Robe, Scipio Society of Naval and Military History; Dr. Robert R. Rofen, Aquatic Research Institute; Tim Ross, U.S. Ski Coaches Association; Professor Mary H. Ross, Virginia Polytechnic Institute; Rosemary Rushka, American Academy of Ophthalmology.

Micael Saba, Attiyeh Foundation; Gabe Samuels, Yellow Spots;

José Luis Perez Sanchez, Commercial Office of the Embassy of Spain; Ronald A. Schachar, Association for the Advancement of Ophthalmology; Schick Division, Warner-Lambert; Janet Seagle, U.S. Golf Association; William Seitz, Neighborhood Cleaners Association; Gwen Sells, Tall Clubs International; Dale Servetnick, Department of the Treasury; Norman F. Sharp, Cigar Association of America; Anthony H. Siegel, Ametek; Dr. M.S. Silberman; Joan Silverman, Citicorp; Dave Smith, Disney Company; Linda Smith, National Restaurant Association; Sid Smith, National Association of Hosiery Manufacturers; Wayne Smith, Sunbeam Appliance Company; Bruce V. Snow, Dairylea Cooperative; Dona Sorensen, Fleischmann's Yeast; Marshall Sorkin, Carter-Wallace; Richard Spader, American Angus Association; Dr. Bob Spanyer, American College of Physicians; John J. Suarez, National Pest Control Association; Amy Sudol, Chase Manhattan; Richard J. Sullivan, Olive Oil Group.

David Taylor, Bank Administration Institute; Thomas A. Tervo, Stearns and Foster Bedding; William D. Toohey, Tobacco Institute; Victor Toth, Multi-Tenant Telecommunications Association; Bob Toy, Telephone Pioneers of America; Jim Trdinich, National League.

Ralph E. Venk, Photographic Society of America; Dennis Vetock, U.S. Army Military History Institute; Elaine Viets, St. Louis *Post-Dispatch*; Gerald F. Voigt, American Concrete Pavement Association; Vulcan Foundry.

Al Wagner, AFC Computer Services; Debbie Walsh, American Federation of Teachers; Belinda Baxter Walsh, Procter & Gamble; Spider Webb, Tattoo Club of America; Monique Wegener, Lenders Bagel Baker; Richard H. Welsh, Jr., Cannon Mills; S.S. White Industrial Products, Pennwalt Corporation; Melvin T. Wilczynski, Lane Drug Company; Dr. Elizabeth Williams, Wyoming State Veterinary Laboratory; Dr. Jack Wilmore, University of Texas; Frank C. Wilson, American Orthopedic Association; Donald W. Wilson, U.S. Ski Educational Foundation; Jerry Wiseman, Atlantic Gelatin; Dr. Robert M. Wold, College of Optometrists in Vision Development; Merry Wooten, Astronomical League; World Impex Bowling.

S.G. Yasinitsky, Orders and Medals Society of America.

Dr. E. Zander, Winthrop Consumer Products; Linda Zirbes, Hyatt Hotels Corporation; Jim Zuckerman, Associated Photographers International.

And to my sources who, for whatever reason, preferred to remain anonymous, thanks for your contribution.

ACKNOWLEDGMENTS

Index

Ice cubes, cloudy versus clear, 106–107

Irons, permanent press settings on, 186–187

Jell-O, fruit in, 149–150
Jet lag, birds and, 33–34
Juicy Fruit gum, flavors in, 242

Ketchup bottles, restaurants mating of, 200
Keys
  automobile, 141–142
  to cities, 99
Kiwifruit in gelatin, 149–150
Kneading and bread, 144–145

Legal-size paper, 197
Letters in alphabet soup, 118–119
License plates on trucks, 96
Light bulbs
  fluorescent, stroking of, 131
  loosening of, 93–94
  in traffic signals, 31–32
Lips, black, on dogs, 38–39
Lunula, fingernails and, 241

Mail
  first class versus priority, 166–167
  translation of foreign, 133
Mail-order ads, six-to-eight week delivery promises of, 70–73
Manhole covers, round shape of, 191
Martinizing, One Hour, 28–29
Mayors, keys to cities and, 99
Medicine bottles, cotton in, 89–90
Memorial Day, Civil War, 168–169
Mickey Mouse, four fingers of, 32
Military salutes, 147–149
Minting of new coins, timing of, 126–128
Money, U.S.
  color of, 83–84
  stars on, 180–182
Monkeys, hair picking of, 26–27
Moon following cars, 18–19
Mountains, hills versus, 97–98
Murder scenes, chalk outlines at, 11–12

National Geographics, saving of, 199
Navy and Army, Captain rank of, 48–50

9 as number to get outside line on telephones, 75–76
98.6 degrees Fahrenheit as comfortable ambient temperature, 240
Noses, clogged nostrils and, 20–21
Nostrils, clogged, 20–21
Numbers, Arabic, 16–17

Oil, automotive, grades of, 182–183
Olive oil, virgin, 174–175
Olives, pitting of, 94–95
One Hour Martinizing, 28–29

Pain, effect of warmth upon, 134–135
Paper, legal-size, 197
PBX systems, 75–76
Pencils
  color of, 108
  numbering of, 109
Pennies, vending machines and, 54–56
Periods in telegrams, 77–78
Permanent press settings on irons, 186–187
Permanents, pregnancy and, 170–171
Physical exams, back tapping and, 145–146
Pineapple in gelatin, 149–150
Plumbing, sound of running water and, 239–240
Poison ivy, grazing animals and, 86–87
Police dogs, urination and defecation of, 67–68
Popcorn, other corns versus, 142–143
Post office, translation of foreign mail and, 133
Potato skins, 12–13
Pregnancy, permanents and, 170–171
Priority mail, first class versus, 166–167
Pubic hair, purpose of, 242–243
Punts, measurement of, 124–125

Queen-size sheets, 87–88

Railroads, width of standard gauges of, 157–159
Ranchers, hanging of boots on fence posts by, 243–245
Ribbons, blue, 57–58
Rice Krispies, noise of, 165

Roaches, death position of, 133–134
Rolls, coldness of airline, 52–53
Roman numerals, calculations with, 105–106
Roosters, crowing and, 3

Saloon doors in Old West, 198
Salutes, military, 147–149
Secretary as U.S. government department head designation, 121–122
Shaving of armpits, 226–229
Sheets, queen-size, 87–88
Shirts
  button arrangement of, 207–209
  starch on, 118
Shoes, single, found on highways, 201–207
Sidewalks, cracks on, 176–178
Silos, shape of, 73–74
Ski poles, downhill, 69
Snack foods and prepricing, 79–80
"Snap! Crackle! and Pop!" of Rice Krispies, 165
Sneezing and eye closure, 84–85
Snow and cold weather, 38
Social security numbers, 91–92
Socks, angle of, 114–115
Sour cream, expiration date on, 132
Stamp pads and moisture retention, 24
Starch on shirts, 118
Stock prices as quoted in eighths of a dollar, 112–113
STOP in telegrams, 76–77
String cheese, 155
Sugar, clumping together of, 103–104
Summer, first day of, 139–141

Tall people, aging and, 229–231
Taste, sense of, in children versus adults, 199
Tattoos, color of, 157
Telegrams
  exclamation marks and, 76–77
  periods and, 77–78
Telephones
  dialing 9 to get outside line, 75–76
  holes in mouthpiece, 14–15
Telephone cords, twisting of, 45
Telescopes, inverted images of, 50–51
Tennis balls and fuzz, 35–36

Television commercials, loudness of, 81–83
Throat, uvula, purpose of, 129
Throwing, sex differences in, 42–44
Tickets, red carbons on airline, 179–180
Toasters, one slice slot on, 183–185
Toenails, growth of, 123
Toilet paper, folding over in hotel bathrooms, 4
Toilet seats in public restrooms, 137–138
Tongues, sticking out of, 199
Toothpicks, round versus flat, 237–238
Toques, purpose of, 66–67
Towels, smelly, 24
Traffic signal light bulbs, 31–32
Treasury, printing of new bills by, 126–128
Trucks, license plates on, 96
Tupperware and home parties, 25–26
Turkeys
  beards on, 99
  white versus dark meat, 53–54
20-20 vision, 143

Uvula, purpose of, 129

Vending machines
  half dollars and, 56–57
  pennies and, 54–56
Videotape versus audio tape, 136–137
Virgin olive oil, 174–175
Vision, 20-20, 143

Warmth, its effect on pain and, 134–135
Western Union
  exclamation marks and, 76–77
  periods and, 77–78
Whistling at sporting events, 199
Winter, first day of, 139–141
Worms as fish food, 110–112

Yawning, contagiousness of, 213–217
Yeast in bread, 144–145
Yellow Pages, advertisements in, 60–63

ZIP code, addresses on envelopes and, 44

Master Index of
Imponderability

Following is a complete index of all ten Imponderables® books and *Who Put the Butter in Butterfly?* The bold number before the colon indicates the book title (see the Title Key below); the numbers that follow the colon are the page numbers. Simple as that.

Title Key

"1/2," meaning of, in street addresses, **1**:156

1040 form, numbering scheme of, **4**:9–10

20-20 vision, origins of, **3**:143

24-second clock in NBA basketball, **1**:29–31

3-D glasses and movies, **5**:151–152

501, Levi jeans, origin of name, **6**:61

7-11 Stores, lack of locks on, **4**:123

7UP, origins of name, **5**:9–10

"86," origins of term, **10**:265; **11**:101–102

9 as number to get outside line on telephones, **3**:75–76

911 as emergency telephone number, **5**:145–146

98.6 degrees Fahrenheit as comfortable ambient temperature, **3**:240

Absorbine Jr., origins of, **8**:47–48

Absorbine Sr., identity of, **8**:47–48

AC, and DC, **2**:21–22

Accents, American, of foreign singers, **4**:125–126

Aches, **3**:104–105

Acre, origins of, **2**:89

Acrylic, virgin, **7**:97–98

Actors, speed of speech of, **5**:203; **6**:241–243

Address labels on magazines, **5**:5

Addresses and ZIP codes, on envelopes, **3**:44

Addresses, street, half-numbers on, **8**:253

Advertising sales in TV programming, overrun live programming and, **1**:50–52

Aerosol cans, shaking of, **3**:178

After-shave lotion, stinging and, **6**:161–162

Age of majority, 21 as, **7**:50–51

Aging and effect on voices, **6**:24–25

Agitators in washing machines, movement of, **4**:56

Air mail and "Par Avion," **8**:39

Air temperature, versus water temperature, perception of, **4**:184

Airlines

chime signals on, **7**:6–8

cold rolls at meal service, **3**:52–53

dearth of female pilots on, **1**:131–133

honey roasted peanuts, **4**:13–14

inflation of oxygen masks, **5**:196–198

tickets, red carbons on, **3**:179–180

Airplanes

coffee bags in lavatories, **4**:64–65

dimming of interior lights, **8**:24–25; **9**:296–297

ear popping and, **2**:130–132

feet swelling and, **2**:31–32

numbering of Boeing jets, **4**:30–31; **5**:235; **8**:251–252; **10**:269–270

"Qantas," spelling of, **8**:134–135

red and green lights on, **4**:152–153

Airplanes (*cont.*)
  seat belts, **8**:141–142
  shoulder harnesses, **8**:141–142;
    9:296
  U.S. flags on exterior, **8**:36–
    37
Alarm clocks, life before, **9**:70–
  74
Alcohol
  in cough medicine, **3**:166
  proof of, **2**:177
Algebra, X in, **9**:131–132
"All wool and a yard wide," origins
  of term, **11**:2
"Allemande," **11**:31
"Alligator" shirts, **9**:297–298
Alphabet, order of, **1**:193–198
Alphabet soup
  distribution of letters, **3**:118–
    119
  outside of U.S., **10**:73
Aluminum cans, crushability of,
  7:157–159
Aluminum foil
  and heat to touch, **8**:145–146
  two sides of, **2**:102
Aluminum foil , on neck of cham-
    pagne bottles, **4**:160–161
Ambidexterity in lobsters, **6**:3–4
Ambulances, snake emblems on,
  6:144–145; **7**:239–240
American accents of foreign
    singers, **4**:125–126
American singles, Kraft, milk in,
  1:247–249
"Ampersand," **11**:86
Amputees, phantom limb sensa-
    tions in, **1**:73–75
Anchors, submarines and, **4**:40–
  41
Angel food cake and position
    while cooling, **7**:43–44
Animal tamers and kitchen chairs,
  7:7–11

Ants
  separation from colony, **6**:44–
    45
  sidewalks and, **2**:37–38
"Apache dance," **11**:30
Apes, hair picking of, **3**:26–27
Appendix, function of, **5**:152–153
Apples, as gifts for teachers,
    2:238; **3**:218–220
Apples and pears, discoloration of,
    4:171
Apples in roasted pigs' mouths,
    10:274–275
April 15, as due date for taxes,
    5:26–29
Aquariums, fear of fish in, **4**:16–
  18
Arabic numbers, origins of, **3**:16–
  17
Architectural pencils, grades of,
    7:73
Area codes, middle digits of,
    5:68–69; **9**:287
Armies, marching patterns of,
    8:264
Armpits, shaving of, **2**:239; **3**:226–
    229; **6**:249
Army and Navy, Captain rank in,
    3:48–50
Art pencils, grades of, **7**:73
Aspirin
  headaches and, **7**:100–102
  safety cap on 100–count bottles
    of, **4**:62
Astrology, different dates for signs
    in, **4**:27–28
Astronauts and itching, **9**:208–216
"At loggerheads," **11**:104–105
Athletics, Oakland, and elephant
    insignia, **6**:14–15
"Atlas," **11**:154–155
ATMs
  swiping of credit cards in,
    10:138–141

swiping versus dipping of credit cards in, **10**:141–142

transaction costs of, **3**:102–103

"Attorney-at-law," **11**:103–104

Auctioneers, chanting of, **9**:201–204

Audiocassette tape on side of road, **7**:250–251

Audiocassette tapes on roadsides, **9**:300

Audiotape, versus videotape technology, **3**:136–137

Audiotape recorders

counter numbers on, **4**:148–149

play and record switches on, **5**:23–24

Automobiles

batteries and concrete floors, **10**:234–236

bright/dimmer switch position, **7**:44–46; **8**:258

bunching of, on highways, **7**:247

cardboard on grills, **6**:188–189; **7**:245–246

clicking sound of turn signals, **6**:203

cockroaches in, **7**:3–4; **8**:256–257; **9**:298

Corvette, 1983, **9**:137–139

cruise controls, **8**:124–125

day/night switch on rear-view mirrors, **4**:185–186

dimples on headlamps, **8**:56

elimination of side vents, **6**:13–14

gas gauges in, **3**:5–6;**6**:273

headlamp shutoff, **7**:92–93

headlights and deer, **6**:212–214

holes in ceiling of, **5**:179

key release button on, **5**:169

keys, ignition and door, **3**:141–142

"new car smell," **5**:63

oil loss after oil change, **7**:240–241

oil, grades of, **3**:182–183

rear windows of, **5**:143–144

rentals, cost in Florida of, **4**:24–25

side-view mirrors, **2**:38–39

speed limits, **2**:143

speedometers, **2**:144–145

tire tread, **2**:72–74

weight of batteries, **5**:101–102

white wall tires, **2**:149

windshield wipers, versus buses, **7**:28

Autopsies of executed criminals, **8**:13–15

"Ax to grind," **11**:70

Babies

blinking, **6**:158–159

burping, **10**:123–124

hair color, **10**:209

high temperatures, tolerance of, **4**:103–104

sleep, **6**:56–57

Baby corn, in supermarkets, **10**:186–187

Baby pigeons, elusiveness of, **1**:254; **10**:251–253

Baby Ruth, origins of name, **8**:84; **9**:288–289; **10**:264–265

Baby shrimp, peeling and cleaning of, **5**:127

"Back and fill," **11**:2–3

Back tapping during physical exams, **3**:145–146

Backlogs in repair shops, **4**:45–47

Bad breath, upon awakening, in the morning, **4**:52

Badges, marshals' and sheriffs', **5**:73–74

Bagels, holes in, **3**:90–91; **8**:261

Bags under eyes, **3**:151

Bags, paper, jagged edges and,
6:117–118
Baked goods
Pennsylvania and, 2:121–122
seven-layer cakes, 6:80–81
unctuous taste until cooled,
6:151–152
Baked potatoes in steak houses,
6:127–129
Ball throwing, sex differences and,
3:42–44
Ballet, "on pointe" in, 8:69–72
Balls, orange, on power lines,
4:18–19
Balls, tennis, fuzz and, 3:35–36
Balsa wood, classification as hard-
wood, 5:85–86
Banana peels as slipping agents,
3:198; 5:228–229; 6:250–252
Bananas, growth of, 2:81–82
Band-Aid packages, red tear
strings on, 1:147–149; 6:266
Bands, tardiness of, in nightclubs,
8:184; 9:248–254
Bands, marching, formation of,
8:107–108
Bands, paper, around Christmas
card envelopes, 6:203–204
Banking
ATM charges, 3:102–103
hours, 3:100–101
Barbecue grills, shape of, 10:99–
102
"Barbecue," 11:173
Barbie, hair of, versus Ken's, 7:4–
5; 8:259–260
Barefoot kickers in football,
4:190–191
Bark, tree, color of, 6:78–79
Barns, red color of, 3:189–191
Bars
mirrors in, 10:14–17
sawdust on floor of, 10:118–120
television sound and, 10:12–14

Baseball
black stripes on bats, 8:104–106
Candlestick Park, origins of,
9:48–51
cap buttons, 9:171–172
caps, green undersides of,
9:172–173
circle next to batter's box in,
3:28
dugout heights, 5:14
first basemen, ball custody of,
1:43–44
home plate sweeping, by um-
pires, 8:27–31
home plate, shape of, 5:131
Japanese uniforms, 10:207–208
"K," as symbol for strikeout,
5:52–53
pitcher's mounds, location of,
5:181; 9:195–198
Baseball cards
wax on wrappers, 5:123
white stuff on gum, 5:122
Basements, lack of, in southern
houses, 4:98
Basketball
24–second clock in NBA, 1:29–
31
duration of periods in, 9:65–69
Basketballs, fake seams on, 4:155–
156
Baskin-Robbins, cost of cones versus cups at, 1:133–135
"Batfowling," 11:1–2
Bathrooms
group visits by females to,
7:183–192; 8:237–238;
9:277–278
ice in urinals of, 10:232–234
in supermarkets, 6:157
Bathtub drains, location of,
3:159–160
Bathubs, overflow mechanisms on,
2:214–215

Bats, baseball, stripes on, **8**:104–106

Batteries
automobile, weight of, **5**:101–102
concrete floors and, **10**:232–234
drainage of, in cassette players, **10**:259–260
nine-volt, shape of, **6**:104; **7**:242–243
sizes of, **3**:116
volume and power loss in, **2**:76–77

"Battle royal," **11**:67

Bazooka Joe, eye patch of, **5**:121

Beacons, police car, colors on, **7**:135–137

"Bead," "Draw a," origins of term, **10**:168

Beaks versus bills, birds and, **10**:3–4

Beanbag packs in electronics boxes, **6**:201

Beans, green, "French" style, **10**:125–126

Beards on turkeys, **3**:99

"Bears [stock market]," **11**:106–107

"Beating around the bush," **11**:1–2

Beavers, dam building and, **10**:42–46

"Bedlam," **11**:69

Beds, mattresses, floral graphics on, **9**:1–2

Beef, red color of, **8**:160–161

Beeps before network news on radio, **1**:166–167

Beer
and plastic bottles, **7**:161–162
steins, lids of, **9**:95–96
temperature in Old West, **5**:17–18

twistoff bottle caps, merits of, **4**:145

Beetle, Volkswagen, elimination of, **2**:192–194

Bell bottoms, sailors and, **2**:84–85

Bells in movie theaters, **1**:88–89

Belly dancers, amplitude of, **5**:202–203; **6**:237–239

Belts, color of, in martial arts, **9**:119–123

Ben-Gay, creator of, **6**:46–47

"Berserk," **11**:68–69

Best Foods Mayonnaise, versus Hellmann's, **1**:211–214

Beverly Hills, "Beverly" in, **8**:16–17

"Beyond the pale," **11**:3

Bias on audiotape, **4**:153–154

Bibles, courtrooms and, **3**:39–41

Bicycles
clicking noises, **5**:10–11
crossbars on, **2**:90–91
tires, **2**:224–226

Bill posting at construction sites, **8**:185; **9**:268–270

Billboards, spinning blades on, **9**:61–64

Bill-counting machines, **6**:271–272

Bills versus beaks, birds and, **10**:3–4

"Bimonthly," **11**:194

"Bingo," origin of term, **5**:202; **6**:231–233; **7**:233–234

Binoculars, adjustments for, **5**:124–125

Bird droppings, color of, **3**:241

"Birdie," **11**:139–140

Birds [see also specific types]
bills versus beaks in, **10**:3–4
coloring of, **5**:11–12
death of, **2**:120–121; **6**:268
direction of takeoff, **7**:107
droppings, color of, **2**:83; **3**:241

Birds (*cont.*)
honking during migration, **7**:108
jet lag and, **3**:33–34
migration patterns of, **5**:147–148; **9**:91–94
sleeping habits of, **5**:72–73
telephone wires and, **3**:130–131
walking versus hopping, **3**:154
worms and, **10**:65–72
Birthday candles, trick, **4**:176
"Birthday suit," **11**:122
Biting of fingernails, reasons for, **9**:223–228
Bitings in keys, **8**:59–60
"Bitter end," **11**:72
"Biweekly," **11**:194
Black
clothing and bohemians, **8**:218–221
gondola color, **4**:86–87
judges' robes, **6**:190–192
specks in ice cream, **8**:132–133
stripes on baseball bats, **8**:104–106
"Blackmail," **11**:188
Blacktop roads, lightening of, **5**:22–23
Blades, razor, hotel slots for, **2**:113
Bleach, flour and, **3**:63–64
Blind
money counting, **3**:152–153
wearing dark glasses, **7**:93–94
Blinking and babies, **6**:158–159
"Blockhead," **11**:80–81
Blood, color of, **4**:138
Blouses, women's, lack of sleeve length sizes in, **4**:113–114
Blue
as color for baby boys, **1**:29
as color of blueprints, **4**:102–103
as color of racquetballs, **4**:8–9
in hair rinses, **2**:117–118

"Blue blood," origin of term, **2**:91
Blue Demons and De Paul University, **8**:76
Blue jeans and orange thread, **9**:74
"Blue jeans," **11**:124
"Blue moon," **11**:12
Blue plate special, origins of, **4**:198; **5**:211–213; **6**:256–257; **7**:227–228; **8**:232–233
Blue ribbons (as first prize), **3**:57–58
"Blue streak," **11**:90
Blueprints, color of, **4**:102–103
"Blurb," **11**:35–36
Blushing and feelings of warmth, **5**:113–114
Boats, red and green lights on, **4**:152–153
"Bobbies [London officers]," **11**:108; **11**:166
"Bobby pins," **11**:167
Body hair
loss of, in humans, **2**:6–8
pubic hair, purpose of, **2**:146
underarm hair, purpose of, **2**:146
Body odor, human immunity to own, **2**:92
Boeing, numbering of jets, **4**:30–31; **5**:235; **8**:251–252; **10**:269–270
"Bogey," **11**:139–140
Boiling water during home births, **6**:114–115; **7**:247–248; **8**:254
Bombs and mushroom clouds, **9**:8–9
Book staining, of paperback book pages, **2**:93–94
Books
checkered cloth on, **7**:126–127
pagination in, **1**:141–144
Boots on ranchers' fence posts, **2**:77–81; **3**:243–245; **4**:231;

5:247–248; 6:268; 7:250;
8:253; 9:298–299; 10:251–
253

"Booze," 11:173

Bottle caps, beer, twistoff, merits
of, 4:145

Bottles, cola, one- and two-liter,
4:188

Bottles, soda, holes on bottom of,
6:187–188

Boulder, Colorado, magazine sub-
scriptions and, 4:33–34

Bowling, "turkey" in, 4:48–49

Bowling pins, stripes on, 6:181–
182

Bowling shoes, rented, ugliness of,
3:173–174

Boxer shorts, versus briefs, 6:11–
12

Boxers, sniffing and, 2:22–23;
10:256–258

Boxes, Japanese, yellow color of,
7:130–131

Bracelets, migration of clasps on,
7:180; 8:197–201; 9:279–
281

Brain, Ten percent use of, alleged,
4:198; 5:210–211; 6:254–
256; 8:232

"Brass tacks," 11:6

"Bravo Zulu," origins of, 8:72–73

Brazil nuts in assortments, 7:145–
147

Bread
"French" versus "Italian,"
7:163–166; 8:261–262
kneading of, 3:144–145
staling of, 7:125–126
yeast in, 3:144–145

"Break a leg," 11:32–33

Breath, bad, upon awakening, in
the morning, 4:52

Brick shape of gold bullion, 8:32–
33

Bricks
holes in, 6:61–62
in skyscrapers, 6:102–103

Bridges versus roads, freezing
characteristics of, 5:193–
194

Bridges, covered, purpose of,
7:132–133; 9:296

Briefs, versus boxer shorts, 6:11–
12

Brights/dimmer switch, position
of, in automobiles, 7:44–46

Broccoli and cans, 8:74–76

Brominated vegetable oil in orange soda, 6:96–97

Brushes, hair, length of handles
on, 7:38–39

Bubble gum
baseball card gum, 5:122
Bazooka Joe's eyepatch, 5:121
bubble-making ingredients,
5:120
flavors of, 5:119–120
pink color of, 3:30–31

"Buck," 11:107

Buckles, pilgrim hats and, 10:47–
51

Buffets, new plates at, 9:5–6;
10:274

Bulbs, light
and noise when shaking, 5:46–
47
halogen, 5:164
high cost of 25–watt variety,
5:91

Bull rings, purpose of, 10:147–
148

"Bulls [stock market]," 11:106–
107

"Bunkum," 11:69–70

Bunnies, chocolate Easter, 2:116

Buns
hamburger bottoms, thinnness
of, 2:32–34

wrapping of boxed chocolate,
8:122–123

"Cannot complete your call as dialed" telephone message,
4:129–130

Cans, aluminum, crushability of,
7:157–159

"Can't hold a candle," 11:71

Capital punishment, hours of execution and, 2:34–36

"Capital punishment," 11:108

"Capitalization," 11:108

Caps and gowns, at graduations,
6:99–102

Caps, baseball
buttons atop, 9:171–172
green undersides, 9:172–173

Captain, Navy and Army rank of,
3:48–50

Caramels, versus toffee, 1:64

Cardboard on automobile grills,
6:188–189

Carpenter's pencils, shape of,
7:27; 9:290–291

CAR-RT SORT on envelopes,
5:78–79

Cars (see "Automobiles"),

Cartoon characters
Casper the ghost, 10:246–249
Donald Duck, 3:150
Goofy, 3:64–65; 7:102
Mickey Mouse, 3:32

Cash register receipts, color of,
4:143

Cashews, lack of shells in, 1:177;
5:235–236

Caskets, position of heads in, 6:8–9

Casper the ghost, identity of,
10:246–249

Cassette players, audio, battery
drainage in, 10:259–260

Cassette tapes, audio, on roadsides, 7:250–251; 9:300

Cat feces, as food for dogs, 6:35–37

Cat food, tuna cans of, 8:8–10

"Cat out of the bag," 11:25

"Catercorner," 11:197

Catnip and effect on wild cats,
6:183–185

Cats
calico, gender of, 6:131–132
catnip, 6:183–185
ear scratching, 5:48–49
eating posture, 6:63–64
hair, stickiness of, 5:163–164
miniature, 6:154–155
sight, during darkness, 1:200–201
swimming and, 1:86

"Cats and dog, raining," 11:26

"Catsup," 11:177

Cattle guards, effectiveness of,
3:115

"Cattycorner," 11:197

Cavities, dogs and, 10:277–278

CDs, Tuesday release of, 10:108–112

Ceiling fans
direction of blades, 9:113–114
dust and, 9:111–113; 10:263–264

Ceilings of train stations, 8:66–67

Celery in restaurant salads, 6:218;
7:207–209; 8:249

Cement
laying of, 8:258
versus concrete, 9:295

Cemeteries
financial strategies of owners of,
2:95–99
perpetual care and, 2:221–222

Ceramic tiles in tunnels, 6:135–136; 8:257–258

Cereal
calorie count of, 1:38–40

Cereal (*cont.*)
joining of flakes in bowl, **8**:115–119
Snap! And Rice Krispies, **8**:2–4
Chalk outlines of murder victims, **3**:11–12
Champagne
aluminum foil on neck of, **4**:160–161
name of, versus sparkling wine, **1**:232–234
Channel 1, lack of, on televisions, **4**:124; **7**:242; **10**:267–268
Chariots, Roman, flimsiness of, **10**:105–107
"Checkmate," **11**:133
Checks
approval in supermarkets, **7**:210–217; **8**:245–246
numbering scheme of, **4**:38–39; **5**:236–237
out-of-state acceptance of, **8**:120–121
white paper attachments, **6**:124–125; **7**:245
Checks, canceled
numbers on, **6**:123
returned, numerical ordering of, **6**:165–166
white paper attachments, **6**:124–125; **7**:245
Cheddar cheese, orange color of, **3**:27–28; **10**:275
Cheese
American, milk in Kraft, **1**:247–249
cheddar, orange color of, **3**:27–28; **10**:275
string, characteristics of, **3**:155
Swiss, holes in, **1**:192
Swiss, slice sizes of, **9**:142–146
Chef's hat, purpose of, **3**:66–67
Chewing gum
lasting flavor, **5**:195–196

water consumption and hardening of, **10**:236–237
wrapping of, **8**:111–112
Chewing motion in elderly people, **7**:79–80
Chianti and straw-covered bottles, **8**:33–35
Chicken
cooking time of, **1**:119–121
versus egg, **4**:128
white meat versus dark meat, **3**:53–54
"Chicken tetrazzini," **11**:153
Children, starving, and bloated stomachs, **7**:149–150
Children's reaction to gifts, **8**:184; **9**:234–237
Chime signals on airlines, **7**:6–8
Chirping of crickets, at night, **10**:54–57
Chocolate
Easter bunnies, **2**:116
shapes of, **2**:24–25
white versus brown, **5**:134–135
wrapping of boxed, **8**:122–123
Chocolate milk, consistency of, **3**:122–123
"Chops," **11**:47
Chopsticks, origins of, **4**:12–13
"Chowderhead," **11**:72
Christmas card envelopes, bands around, **6**:203–204
Christmas tree lights
burnout of, **6**:65–66
lack of purple bulbs in, **6**:185–186
purple, **9**:293; **10**:280
Cigar bands, function of, **4**:54–55
Cigarette butts, burning of, **5**:45
Cigarettes
grading, **6**:112
odor of first puff, **2**:238; **3**:223–226
spots on filters, **6**:112–113

Cigars, new fathers and, **3**:21–22

Cities, higher temperatures in, compared to outlying areas, **1**:168–169

Civil War, commemoration of, **3**:168–169

"Claptrap," **11**:73

Clasps, migration of necklace and bracelet, **7**:180; **8**:197–201; **9**:279–281

Cleansers, "industrial use" versus "household," **5**:64–65

Clef, treble, dots on, **10**:210–213

Clicking noise of turn signals, in automobiles, **6**:203

Climate, West coast versus East coast, **4**:174–175

Clocks
clockwise movement of, **2**:150
grandfather, **4**:178
number 4 on, **2**:151–152
Roman numerals, **5**:237–238
school, backward clicking of minute hands in, **1**:178–179
versus watches, distinctions between, **4**:77–78

Clockwise, draining, south of the Equator, **4**:124–125

"Cloud nine," **11**:97–98

Clouds
disappearance of, **5**:154
location of, **3**:13
rain and darkness of, **2**:152

Clouds in tap water, **9**:126–127

"Cob/cobweb," **11**:20

Coca-Cola
2–liter bottles, **5**:243–244
taste of different size containers, **2**:157–159

Cockroaches
automobiles and, **7**:34; **8**:256–257; **9**:298

death position of, **3**:133–134; **8**:256

reaction to light, **6**:20–21

Coffee
bags in lavatories of airplanes, **4**:64–65
bags versus cans, **4**:146
electric drip versus electric perk in, **4**:35
restaurant versus home-brewed, **7**:181; **8**:221

Coffee, decaffeinated
leftover caffeine usage, **6**:195
orange pots in restaurants, **6**:67–69

Coffeemakers, automatic drip, cold water and, **4**:173

Coins
lettering on U.S. pennies, **7**:5
red paint on, **7**:117
serrated edges of, **1**:40–41
smooth edges of, **1**:40–41

Cola bottles, one- and two-liter, black bottom of, **4**:188

Colas, carbonation in, **1**:87–88

Cold water and automatic drip coffeemakers, **4**:173

Colds
body aches and, **3**:104–105
clogged nostrils and, **3**:20–21
liquids as treatment for, **4**:131–132
symptoms at night of, **3**:163

"Coleslaw," **11**:174

"Collins, Tom [drink]," **11**:165

Color
blood, **4**:138
cash register receipts, **4**:143
wet things, **4**:139

Comic strips and capital letters, **5**:55–56

Commercials, television, loudness of, **3**:81–83

Diet soft drinks
  calorie constituents in, **6**:94–95
  phenylalanine as ingredient in,
    6:96
Dimples
  auto headlamps, **8**:56
  facial, **3**:23
  golf balls, **3**:45–46
Dinner knives, rounded edges of,
    1:231–232
Dinner plates
  repositioning of, **8**:184
  round shape of, **8**:162–164
Dirt, refilling of, in holes, **7**:48–49
Disc jockeys and lack of song
    identification, **7**:51–57
"Discussing Uganda," origins of
    term, **5**:246–247; **11**:145
Dishwashers, two compartments
    of, **6**:109–110
Disney cartoon characters
  Donald Duck, **3**:150
  Goofy, **3**:64–65; **7**:102
  Mickey Mouse, **3**:32
Disposable lighters, fluid cham-
    bers of, **6**:92–93
Distilleries, liquor, during Prohibi-
    tion, **9**:54–56
Ditto masters, color of, **6**:133–134
"Dixie," **11**:147–148
Dixieland music at political rallies,
    5:203; **6**:234–236
"Dixieland," **11**:147–148
DNA, identical twins and, **10**:18–
    20
Doctors, back tapping of, **3**:145–
    146
Doctors and bad penmanship,
    5:201; **6**:221–225; **7**:232–
    233; **8**:235
"Dog days," **11**:17–18
Dogs
  barking, laryngitis in, **2**:53–54
  black lips of, **3**:38–39

body odor of, **2**:40
cavities and, **10**:277–278
circling before lying down, **2**:2–
    3; **5**:238–239
crooked back legs of, **8**:126–
    128
Dalmatians and firefighting,
    6:270–271
drooling in, **6**:34–35
eating cat feces, **6**:35–37
eating posture of, **6**:63–64
head tilting of, **4**:198; **5**:215–
    217; **6**:258–259; **8**:237
lifting legs to urinate in, **4**:35–
    36
miniature, **6**:154–155
poodles, wild, **6**:207–209
rear-leg wiggling, when
    scratched, **6**:52–53
"sic" command, **5**:51
sticking head out of car win-
    dows in, **4**:60–61
wet noses, **4**:70–73
Dollar sign, origins of, **7**:103–104
Dolls, hair of, **7**:4–5
Donald Duck, brother of, **3**:150
Donkey Kong, origins of, **10**:38–
    40
"DONT WALK" signs, lack of
    apostrophes in, **6**:75–76
Doors
  double, in stores, **6**:177–180
  knobs versus handles on,
    7:148–149
  opening orientation of, in build-
    ings, **4**:167
  shopping mall entrances,
    6:180–181
  "THIS DOOR TO REMAIN
    UNLOCKED DURING
    BUSINESS HOURS" signs,
    in stores, **6**:202
"Doozy," **11**:37
"Do-re-mi," **11**:29–30

Dots on cue balls, in pool,
10:237–240
Double doors in stores, **6**:177–
180
"Doubleheader," **11**:133
Double-jointedness, **10**:229–231
Double-yolk eggs, **3**:188–189
Doughnuts
tissues and handling in stores,
2:164; **5**:240
origins of holes in, **2**:62–64
Downhill ski poles, shape of,
3:69
Dr Pepper
origins of name, **5**:129–130;
6:272
punctuation of, **8**:253–254
Drains, location of bathtub,
3:159–160
"Draw a Bead," origins of term,
10:168; **11**:134
Dreams, nap versus nighttime,
3:124
Drinking glasses, "sweating" of,
9:124–125; **10**:261
Dripless candles, whereabouts of
wax in, **4**:182–183
"Driveway," **11**:64–65
Driveways, driving on, versus
parkways, **4**:123
Driving, left- versus right-hand
side, **2**:238; **3**:209–212;
6:248–249; **7**:223; **8**:230–
231
Drooling, dogs and, **6**:34–35
Drowsiness after meals, **6**:138–
139
Drugstores, high platforms in,
8:5–7
"Dry," as terms for wines, **5**:141–
142
Dry-cleaning
French, **3**:164–165
garment labels and, **2**:59–60

One Hour Martinizing, **3**:28–29
raincoats and, **2**:216–217
Dryers, hand
in bathrooms, **10**:266–267
"off" switches, **8**:174–176
Ducks
lakes and ponds and, **10**:256
on Cadillacs, **5**:174–176
Duels, timing of, **5**:69
Dugouts, height of, **5**:14
"Dukes," **11**:137
"Dumb [mute]," **11**:131–132
"Dumbbells," **11**:131–132
Dust, ceiling fans and, **10**:263–
264

E
as school grade, **3**:198; **4**:206–
209
on eye charts, **3**:9–10
"Eagle [golf score]," **11**:139–140
Earlobes, function of, **5**:87–88
"Earmark," **11**:46
Earrings, pirates and, **9**:43–45;
10:272–273
Ears
hairy, in old men, **2**:239; **3**:231–
233; **5**:227–228
popping in airplanes, **2**:130–
132
ringing, causes of, **2**:115–116
Earthworms as fish food, **3**:110–
112
Easter
chocolate bunnies and, **2**:116
dates of, **4**:55–56
ham consumption at, **1**:151–152
"Easy as pie," **11**:172
Eating, effect of sleep on, **6**:138–
139
"Eavesdropper," **11**:109–110
Ebert, Roger, versus Gene Siskel,
billing of, **1**:137–139
Egg, versus chicken, **4**:128

Egg whites and copper bowls,
7:99

Eggs
color of, 2:189–190
double-yolk, 3:188–189
hard-boiled, discoloration of,
3:34
meaning of grading of, 4:136–
137
sizes of, 2:186–188

"Eggs Benedict," 11:154

"Eighty-six," origins of term,
10:265–266;
11:101–102

Elbow macaroni, shape of, 4:28

Elderly men
pants height and, 2:171–172
shortness of, 2:239; 3:229–231;
6:250

Elections, U.S.
timing of, 6:41; 8:260–261;
9:291–292
Tuesdays and, 1:52–54; 3:239

Electric can openers, sharpness of
blades on, 6:176–177

Electric drip versus electric perk,
in coffee, 4:35

Electric perk versus electric drip,
in coffee, 4:35

Electric plug prongs
holes at end of, 5:94–95
three prongs versus two prongs,
5:191

Electricity, AC versus DC, 2:21–
22

Electricity, static, variability in
amounts of, 4:105–106

Elephants
disposal of remains of, 6:196–
197
jumping ability of, 10:27–29
Oakland A's uniforms, 6:14–15
rocking in zoos, 8:26–27;
10:279

Elevator doors
changing directions, 8:169–170
holes, 8:170–171

Elevators
overloading of, 5:239
passenger capacity in, 1:23–24

"Eleventh hour," 11:99–100

Emergency Broadcasting System,
length of test, 1:117–118

"Emergency feed" on paper towel
dispensers, 8:149–150

Emmy awards, origins of name,
2:52

English, driving habits of, 2:238;
3:209–212; 6:248–249;
7:223; 8:230–231

English horn, naming of, 5:38–39

English muffins
in England, 6:200–201
white particles on, 6:49

Envelopes
Christmas cards, paper bands
on, 6:203–204
colored stickers on, 4:83–84
computer scrawls on, 5:104–
106
red letters on back of, 5:50
return windows on, 2:111;
5:239–240
size of, in bills, 6:81–83

Escalator knobs, purpose of, 2:42

Escalators
green lights on, 3:172
rail speed of, 3:171

Evaporated milk
refrigeration of, 2:114
soldered cans and, 2:114

Evolution, loss of body hair and,
2:6–8

Exclamation marks in telegrams,
3:76–77

Executions
autopsies after, 8:13–15
hours of, 2:34–36

Fingernails
  biting of, **8**:183; **9**:223–228
  growth of, versus toenails, **3**:123
  growth of, after death, **4**:163–
    164
  lunula on, **2**:218; **3**:241
  yellowing of, and nail polish,
    7:129–130
Fingers, length of, **5**:40–41
"Fink," **11**:73
Fire, crackling sound of, **4**:11
Fire extinguishers, expiration date
  on, **4**:68
Fire helmets, shape of, **10**:199–
  203
Fire hydrants
  freezing of water in, **6**:150;
    10:11
  shape of valve stems of, **4**:142–
    143
Fire trucks, roofs of, **7**:143–144
Firearms, confiscated, after ar-
  rests, **7**:85–87
Fireplaces, rain in, **2**:57–58
First basemen, ball custody of, in
  baseball, **1**:43–44
First class mail, priority mail versus, **3**:166–167
Fish [see also specific types]
  biting in rain, **10**:131–138
  eating preferences, **3**:110–112
  fear of, in aquariums, **4**:16–18
  floating upon death, **5**:186–
    187
  pebble spitting, **9**:174–175
  return to dried-up lakes and
    ponds, **3**:15–16; **10**:256
  saltwater coloration, **9**:133–136
  sleep patterns, **3**:161–162;
    6:270
  urination, **7**:90–91
Fish tanks, oxygen in, **7**:84–85
Flagpoles, balls atop, **2**:17–18;
  6:268

Flags, Japanese, red beams and,
  10:151–155
Flags, U.S.
  half-mast, **10**:36–38
  on airplanes, **8**:36–37
"Flak," **11**:74
"Flammable," versus "inflamma-
  ble," **2**:207–208
"Flash in the pan," **11**:74
Flashlights, police and grips of,
  10:30–32
Flat toothpicks, purpose of,
  1:224–225
"Flea market," **11**:24
Flies
  landing patterns of, **2**:220
  winter and, **2**:133–135
Flintstones, Barney Rubble's pro-
  fession on, **7**:173–174
Flintstones multivitamins, Betty
  Rubble and, **6**:4–5; **9**:285–
  286
Floaters in eyes, **3**:37
Floral graphics on mattresses,
  9:1–2
Florida, cost of auto rentals in,
  4:24–25
"Flotsam," versus "jetsam," **2**:60–
  61
Flour, bleached, **3**:63–64
Flour bugs, provenance of, **4**:89–
  90
Flour tortillas, size of, versus corn,
  10:142–145
Flu, body aches and, **3**:104–105
Fluorescent lights and plinking
  noises, **5**:47
Flush handles, toilets and, **2**:195–
  196
Flushes, loud, toilets in public
  restrooms and, **4**:187
Fly swatters, holes in, **4**:31–32
FM radio, odd frequency numbers
  of, **10**:59–60

Food cravings in pregnant women, **10**:183–185

Food labels
  "FD&C" on label, **4**:163
  lack of manufacturer street addresses, **4**:85

Football
  barefoot kickers in, **4**:190–191
  college, redshirting in, **7**:46–48
  distribution of game balls, **2**:44
  goalposts, tearing down of, **8**:213–217
  measurement of first-down yardage, **5**:128–129
  origins of "hut" in, **9**:294–295
  Pittsburgh Steelers' helmet emblems on, **7**:67–68
  shape of, **4**:79–81
  sideline population in, **10**:51–53
  tearing down of goalposts, **7**:181
  two-minute warning and, **10**:150–151
  yardage of kickers in, **3**:124–125

"Fore," origins of golf expression, **2**:34

Forewords in books, versus introductions and prefaces, **1**:72–73

Forks, switching hands to use, **3**:198

"Fortnight," **11**:194

Fraternities, Greek names of, **10**:94–98

"Freebooter," **11**:110

Freezer compartments, location of, in refrigerators, **2**:230–231

Freezers
  ice trays in, **10**:92–93
  lights in, **10**:82–85

French dry cleaning, **3**:164–165

French horns, design of, **5**:110–111

"French" bread versus "Italian," **8**:261–262

"French" style green beans, **10**:125–126

"French" versus "Italian" bread, **7**:165–166

Frogs
  eye closure when swallowing, **6**:115–116
  warts and, **10**:121–123

Frogs of violins, white dots on, **4**:164–165; **9**:291

Frostbite, penguin feet and, **1**:217–218

Fruitcake, alleged popularity of, **4**:197; **5**:205–209; **6**:253; **7**:226–227; **9**:274–276

"Fry," **11**:179

Fuel gauges in automobiles, **6**:273

"Fullback," **11**:138

Full-service versus self-service, pricing of, at gas stations, **1**:203–209

Funeral homes, size of, **8**:152–155

Funerals
  burials without shoes, **7**:53–54
  depth of graves, **7**:14–15
  head position of, in caskets, **6**:8–9
  orientation of deceased, **7**:106
  perpetual care and, **2**:221–222
  small cemeteries and, **2**:95–99

"Funk," **11**:75

"G.I.," **11**:52

Gagging, hairs in mouth and, **7**:76–77

Gallons and quarts, American versus British, **4**:114–115

Game balls, distribution of football, **2**:44

Gasoline, pricing of, in tenths of a cent, **2**:197–198

Graves, depth of, **7**:14–15

Gravy skin loss, when heated, **6**:58

"Gravy train," **11**:62

Grease, color of, **5**:182

Grecian Formula, process of, **8**:110–111

Greek names of fraternities and sororities, **10**:94–98

Green beans, "French" style, **10**:125–126

Green color of glow-in-the-dark items, **9**:139–141

Green lights, versus red lights, on boats and airplanes, **4**:152–153

"Green with envy," **11**:188

"Green" cards, color of, **7**:61–63

"Greenhorn," **11**:189

Greeting cards, shape of, **6**:70–71

Gretzky, Wayne, hockey uniform of, **2**:18; **10**:279

Grimace, identity of McDonald's, **7**:173

Grocery coupons, cash value of, **5**:7–9

Grocery sacks, names on, **2**:166–167

Grocery stores and check approval, **8**:245–246

Groom, carrying bride over threshold by, **4**:159

Growling of stomach, causes of, **4**:120–121

Guitar strings, dangling of, **8**:11–13; **10**:276

Gulls, sea, in parking lots, **6**:198–199; **10**:254–256

Gum, chewing
water consumption and hardening of, **10**:236–237
wrappers of, **8**:111–112

"Gunny sacks," **11**:195

"Guy," **11**:151–152

"Habit [riding costume]," **11**:123

"Hackles," **11**:6–7

Hail, measurement of, **5**:203; **6**:239–240; **7**:234–235; **8**:236–237

Hair
blue, and older women, **2**:117–118
growth of, after death, **4**:163–164
length of, in older women, **7**:179; **8**:192–197
mole, color of, **8**:167–169
parting, left versus right, **1**:116

Hair color, darkening of, in babies, **10**:209

Hair spray, unscented, smell of, **2**:184

Hairbrushes, length of handles on, **7**:38–39

Hairs in mouth, gagging on, **7**:76–77

Hairy ears in older men, **2**:239; **3**:231–233; **5**:227–228

Half dollars, vending machines and, **3**:54–56

"Halfback," **11**:138

Half-mast, flags at, **10**:36–38

Half-moon versus quarter moon, **7**:72–73

Half-numbers in street addresses, **8**:253

Halibut, coloring of, **3**:95–96

Halloween, Jack-o'-lanterns and, **4**:180–181

Halogen light bulbs, touching of, **5**:164

Ham
checkerboard pattern atop, **7**:66–67
color of, when cooked, **7**:15–16
Easter and consumption of, **1**:151–152

"Ham [actor]," **11**:170–171

Hamburger buns, bottoms of, **2**:32–34

"Hamburger," origins of term, **4**:125

"Hamfatter," **11**:170–171

Hand dryers in bathrooms, **8**:174–176; **10**:266–267

Hand positions in old photographs, **7**:24–26

Handles versus knobs, on doors, **7**:148–149

Handwriting, teaching of cursive versus printing, **7**:34–37

"Hansom cab," **11**:63

Happy endings, crying and, **1**:79–80

Hard hats
   backward positioning of, in ironworkers, **4**:94
   exterminators and, **2**:51

Hard-boiled eggs, discoloration of, **3**:34

Hat tricks, in hockey, **2**:165–166

Hats
   cowboy, dents on, **7**:249–250
   declining popularity, **5**:202; **6**:227–231; **7**:233; **7**:249
   dents in cowboy, **5**:6; **6**:274
   holes in sides of, **5**:126
   numbering system for sizes, **4**:110

Haystacks, shape of, **6**:47–48; **8**:265–266

"Hazard [dice game]," **11**:134

"Head [bathroom]," **11**:48

"Head honcho," **11**:39

Head injuries, "seeing stars" and, **10**:156–158

Head lice, kids and, **10**:225–227

Headaches and aspirin, **7**:100–102

Headbands on books, **7**:126–127

Headlamps, shutoff of automobile, **7**:92–93

"Heart on his sleeve," **11**:128

Hearts, shape of, idealized versus real, **4**:199; **5**:220–221; **6**:260; **7**:229–230; **8**:234; **9**:234

Heat and effect on sleep, **6**:137–138

"Hector," **11**:155

"Heebie jeebies," **11**:40

Height
   clearance signs on highways, **8**:156–158
   of elderly, **6**:250
   restrictions on fences, **2**:28–30
   voice pitch and, **2**:70

Heinz ketchup labels, **8**:150–151

Helicopters, noise of, **8**:164–166

Helium and effect on voice, **5**:108–109

Hellman's Mayonnaise, versus Best Foods, **1**:211–214

"Hem and haw," **11**:195–196

"Hep," **11**:52–53

Hermit crabs, bathroom habits of, **7**:74–75

Hernia exams and "Turn your head and cough," **5**:114–115

"Heroin," **11**:77

"High bias," versus "low bias," on audio tape, **4**:153–154

"High jinks," **11**:41

High-altitude tennis balls, **8**:80

"Highball," **11**:175–176

Highways
   clumping of traffic, **4**:165–167; **7**:247
   curves on, **7**:121–122
   interstate, numbering system, **4**:66–67
   traffic jams, clearing of, **1**:25–26
   weigh stations, **4**:193–194

"Hillbilly," **11**:148

Hills, versus mountains, **3**:97–98; **8**:252

MASTER INDEX OF IMPONDERABILITY

Knives, dinner, rounded edges of, **1**:231–232

Knives, serrated, lack of, in place settings, **4**:109–110

Knobs versus handles, on doors, **7**:148–149

"Knock on wood," **11**:9–10

"Knuckle under," **11**:9–10

Knuckles, wrinkles on, **5**:182–183

Kodak, origins of name, **5**:169–170; **9**:290

Kool-Aid and metal containers, **8**:51

Kraft American cheese, milk in, **1**:247–249

"L.S.," meaning of, in contracts, **1**:165

Label warnings, mattress, **2**:1–2

Labels on underwear, location of, **4**:4–5

Labels, food, lack of manufacturer street addresses on, **4**:85

"Ladybug," **11**:23

Ladybugs, spots on, **7**:39–40

Lakes
    effect of moon on, **5**:138–139
    fish returning to dried, **3**:15–16; **10**:256
    ice formations on, **5**:82–83
    versus ponds, differences between, **5**:29–30; **7**:241
    versus ponds, water level of, **9**:85–86
    wind variations, **4**:156–157

"Lame duck," **11**:24–25

Lane reflectors, fastening of, **5**:98–99

Large-type books, size of, **5**:135

Laryngitis, dogs, barking, and, **2**:53–54

Lasagna, crimped edges of, **5**:61

"Last ditch," **11**:10

"Last straw," **11**:8–9

Laughing hyenas, laughter in, **8**:1–2

Lawn ornaments, plastic deer as, **8**:185

Lawns, reasons for, **2**:47–50

"Lawyer," **11**:103–104

"Lb. [pound]," **11**:56

Leader, film, **2**:9

"Leap year," **11**:193

Leather, high cost of, **8**:21–23

Ledges in buildings, purpose of, **8**:18–20

Left hands, placement of wristwatches on, **4**:134–135; **6**:271

"Left wing," **11**:116

Left-handed string players, in orchestras, **9**:108–109; **10**:276–277

Leg kicking by women while kissing, **6**:218; **7**:196–197; **9**:278; 299–300

Legal-size paper, origins of, **3**:197

"Legitimate" theater, origins of term, **10**:5–7

"Let the cat out of the bag," **11**:25

Letters
    business, format of, **7**:180; **8**:201–204
    compensation for, between countries, **4**:5–6

Letters in alphabet soup, distribution of, **3**:118–119

Levi jeans
    colored tabs, **6**:59–61
    origin of "501" name, **6**:61

Liberal arts, origins of, **5**:70–73

Lice, head, kids and, **10**:225–227

License plates and prisoners, **8**:137–139; **10**:268–269

License plates on trucks, absence of, **3**:98; **10**:270–271

"Licking his chops," **11**:47

Licorice, ridges on, **9**:188–189

Life Savers, wintergreen,

feelings of coldness, **6**:218; **7**:198–199

remote controls and, **6**:217; **7**:193–196

Menstruation, synchronization of, in women, **4**:100–102

Menthol, coolness of, **5**:192

Meter, origins of, **2**:200–202

Miami, University of
football helmets, **8**:171–172
"Hurricanes" nickname, **8**:171–172

Mickey Mouse, four fingers of, **3**:32; **6**:271

Microphones, press conferences and, **2**:11–12

Migration of birds, **9**:91–94

Mile, length of, origins of, **1**:241–242

Military salutes, origins of, **3**:147–149

Milk
as sleep-inducer, **7**:17
fat content in lowfat, **7**:60–61
in refrigerators, coldness of, **5**:4–5
Indianapolis 500, **8**:130–131
national brands, **1**:227–231; **9**:287
plastic milk containers, **7**:61; **10**:262–263
single serving cartons of, **7**:137–138
skim versus nonfat, **7**:59
skin on, when heated, **6**:58

Milk cartons
design of, **5**:112; **9**:289–290
difficulty in opening and closing of, **1**:243–246; **5**:243

Milk cases, warnings on, **5**:43–44

Milk Duds, shape of, **8**:81–82

Millimeters, as measurement unit for film, **1**:44

"Mind your P's and Q's," **11**:88–89

Mint flavoring on toothpicks, **4**:153

Mint, U.S., and shipment of coin sets, **5**:32

Minting of new coins, timing of, **3**:128

Mirrors in bars, **10**:14–17

Mirrors, rear-view, **4**:185–186

Mistletoe, kissing under, origins of, **4**:106–107

Mobile homes, tires atop, in trailer parks, **6**:163–164

Mole hair, color of, **8**:167–169

Money, U.S.
color of, **3**:83–84
stars on, **3**:180–182

Monitors, computer, shape of, **6**:129–131

Monkeys, hair picking of, **3**:26–27

Monopoly, playing tokens in, **10**:21–23

Montreal Canadiens, uniforms of, **5**:165; **7**:242

Moon
apparent size of, at horizon, **2**:202–204
effect on lakes and ponds, **5**:138–139
official name, **5**:19–20
quarter-, vs. half-, **7**:72–73

Moons on outhouse doors, **4**:126

Mosquitoes
biting and itching, **5**:3–4
biting preferences, **8**:177–179; **10**:278
daytime habits, **5**:77–78
male versus female eating habits, **5**:190

Moths, reaction to light of, **6**:21–23

Mottoes on sundials, **6**:54–56

Mountains
falling hot air at, **9**:149–151
versus hills, **3**:97–98; **8**:252

Movie actors and speed of speech, **5**:203; **6**:241–243

Movie theaters
bells in, **1**:88–89
in-house popcorn popping, **1**:45–50

Movies, Roman numerals in copyright dates in, **1**:214–216

"Mrs.," **11**:57

MSG, Chinese restaurants and, **2**:168–171

"Mugwump," **11**:119

Muppets, left-handedness of, **7**:111–113

Murder scenes, chalk outlines at, **3**:11–12

Musketeers, Three, lack of muskets of, **7**:29–30

Mustaches, policemen and, **6**:219; **7**:218–220; **8**:246–247; **9**:278

"Muumuu," **11**:125

"Mystery 7," in *$25,000 Pyramid*, **1**:192

Nabisco Saltine packages, red tear strip on, **1**:147–149

Nabisco Shredded Wheat box, Niagara Falls on, **5**:100–101

Nail polish and fingernail yellowing, **7**:129–130

"Namby pamby," **11**:79

National Geographics, saving issues of, **3**:199; **5**:229–230; **7**:224

Navy and Army, Captain rank in, **3**:48–50

Necklaces and clasp migration, **7**:180; **8**:197–201; **9**:279–281

Neckties
direction of stripes on, **6**:86–87
origins of, **4**:127; **8**:264–265
taper of, **6**:84–85

Nectarines, canned, lack of, **4**:59–60; **9**:287–288

Needles, holes in, of syringes, **10**:57–59

Neptune's moon, Triton, orbit pattern of, **4**:117–118

Nerdiness and eyeglasses, **7**:180

"Netherlands," versus "Holland," **2**:65–66

New York City and steam in streets, **5**:16–17

"New York" steaks, origins of, **7**:155–156; **8**:252

New Zealand, versus "Old Zealand," **4**:21–22

Newspapers
ink and recycling of, **7**:139–140
ink smudges on, **2**:209–212
jumps in, **5**:116–117
tearing of, **2**:64
window cleaning and, **10**:33–36
yellowing of, **8**:51–52

Niagara Falls on Nabisco Shredded Wheat box, **5**:100–101

"Nick of time," **11**:158

Nickels, smooth edges of, **1**:40–41

"Nickname," **11**:163

Nightclubs, lateness of bands in, **8**:184; **9**:248–254

"Nine-day wonder," **11**:98

Nine-volt batteries, shape of, **6**:104; **7**:242–243

Nipples, purpose of, in men, **4**:126; **6**:275

"No bones about it," **11**:49

"No Outlet" signs, versus "Dead End" signs, **4**:93

Noise, traffic, U.S. versus foreign countries, **4**:198

North Carolina, University of, and Tar Heels, **8**:76–77

North Pole
directions at, **10**:243
telling time at, **10**:241–243

Nose rings and bulls, **10**:147–148

Noses
clogged nostrils and, **3**:20–21
runny, in cold weather, **10**:146–147
runny, kids versus adults, **9**:89–90
wet, in dogs, **4**:70–73
Nostrils, clogged, **3**:20–21
Notches on bottom of shampoo bottles, **10**:29–30
Notre Dame fighting Irish, **10**:115–117
NPR radio stations, low frequency numbers of, **10**:181–183
Numbers, Arabic, origins of, **3**:16–17
Nutrition labels, statement of fats on, **6**:142–143
Nuts
Brazil, in assortments, **7**:145–147
Macadamia shells, **8**:262
peanuts in plain M&M's, **7**:239
peanuts, and growth in pairs, **7**:34

"O' " in Irish names, **8**:135–136
Oakland A's, elephant on uniforms of, **6**:14–15
Oboes, use of as pitch providers, in orchestras, **4**:26–27
Occupancy, maximum, in public rooms, **10**:158–160
Oceans
boundaries between, **10**:74–76
color of, **2**:213
salt in, **5**:149–150
versus seas, **5**:30–32
Octopus throwing, Detroit Red Wings and, **9**:183–186
"Off the schneider," **11**:136
Oh Henry, origins of name of, **8**:83–84

Oil
automotive, after oil change, **5**:184–185; **7**:240–241
automotive, grades of, **3**:182–183
"Okay," thumbs-up gesture as, **1**:209–210
Oktoberfest, September celebration of, **9**:156–157
Old men
hairy ears and, **2**:239; **3**:231–233; **5**:227–228
pants height and, **2**:171–172; **6**:274
"Old No. 7" and Jack Daniel's, **8**:144–145
"Old Zealand," versus New Zealand, **4**:21–22
Olive Oil, virgin, **3**:174–175
Olives, green and black, containers of, **1**:123–127
"On pointe" and ballet, **8**:69–72
"On tenterhooks," **11**:10–11
"On the Q.T.," **11**:59
"Once in a blue moon," **11**:12
"One fell swoop," **11**:197
One Hour Martinizing, **3**:28–29
One-hour photo processing, length of black-and-white film and, **4**:39
Onions and crying, **9**:169–170
Orange coffee pots, in restaurants, **6**:67–69
Orange juice
price of fresh versus frozen, **5**:155–156
taste of, with toothpaste, **10**:244–246
Orange thread in blue jeans, **9**:74
Oranges, extra wedges of, **4**:175–176
Oranges, mandarin, peeling of, **8**:106–107

"Oreo," origins of name, **2**:173–174

Outhouse doors, moons on, **4**:126

Ovens, thermometers in, **10**:85–87

Overflow mechanism, kitchen sinks and, **2**:214–215

Oxygen in tropical fish tanks, **7**:84–85

Oxygen masks, inflation of airline, **5**:196–198

"P.U.," origins of term, **10**:26–27

"Pacific Ocean," **11**:149

Page numbers on magazines, **4**:14–15

Pagination in books, **1**:141–144

Pain, effect of warmth upon, **3**:134–135

Paint, homes and white, **2**:100–102

Paint, red, on coins, **7**:117

Painters and white uniforms, **6**:17–19

Palms, sunburn on, **8**:63–64

Pandas, double names of, **8**:172–173

Pants, height of old men's, **2**:171–172; **6**:274

"Pantywaist," **11**:81

"Pap test," **11**:49

Paper
legal size, origins of, **3**:197
recycling of holes in loose-leaf, **7**:105–106

Paper cups, shape of, **9**:289

Paper cuts, pain and, **2**:103–104

Paper mills, smell of, **5**:96–98

Paper sacks
jagged edges on, **6**:117–118
names on, **2**:166–167

Paper towel dispensers, "emergency feed" on, **8**:149–150

Paperback books, staining of, **2**:93–94

Papers, yellowing of, **8**:51–52

"Par [golf course]," **11**:139–140

"Par Avion" on air mail postage, **8**:39

"Pardon my French," **11**:150

Parking lots, sea gulls at, **10**:254–256

Parking meters, yellow "violation" flags and, **4**:42–43

"Parkway," **11**:65

Parkways, parking on, versus driveways, **4**:123

Parrots and head bobbing, **8**:23–24

Parting of hair, left versus right, **1**:116

Partly cloudy, versus partly sunny, **1**:21–22

Partly sunny, versus partly cloudy, **1**:21–22

"Pass the buck," **11**:107

Pasta
boxes, numbers on, **4**:107
foaming when boiling, **7**:78
holes in, **4**:28

Pay phones, collection of money from, **1**:107–108

Pay toilets, disappearance of, **2**:25–26

PBX systems, **3**:75–76

"Pea jacket," **11**:124

Peaches
canned, and pear juice, **8**:46–47
fuzziness of, **4**:58–59

Peanut butter, stickiness of, **10**:204–207

Peanuts
allergies to, **7**:239
growth in pairs, **7**:34
honey roasted, and airlines, **4**:13–14
origins of comics name, **10**:191–193

Pear juice in canned peaches, **8**:46–47

Pears and apples, discoloration of, **4**:171

Pebbles, spitting by fish of, **9**:174–175

"Peeping Tom," **11**:162

Pencils
architectural and art, grades of, **7**:173
carpenter's, shape of, **7**:27; **9**:290–291
color, **3**:108
numbering, **3**:109

Penguins
frostbite on feet, **1**:217–218
knees, **5**:160

Penicillin and diet, **8**:95–96

Penmanship of doctors, bad, **5**:201; **6**:221–225; **7**:232–233; **8**:235

Pennies
lettering on, **7**:5
smooth edges of, **1**:40–41
vending machines and, **3**:54–56

Pennsylvania Dept. of Agriculture, registration, baked goods, **2**:121–122

Penny loafers, origins of, **8**:43–44

Pens
disappearance of, **4**:199; **5**:222–223; **6**:260–261; **7**:230–231; **8**:234
holes in barrel of cheap, **4**:111
ink leakages in, **4**:112–113

Pepper
and salt, as condiments, **5**:201; **6**:225–226; **8**:235–236
and sneezing, **8**:61
white, source of, **2**:135–136

Pepsi-Cola, trademark location of, **5**:115–116

Perfume
color of, **9**:19
wrists and, **6**:90

Periods in telegrams, **3**:77–78

Permanent press settings on irons, **3**:186–187

Permanents, pregnancy and, **3**:170–171

Perpetual care, cemeteries and, **2**:221–222

"Peter out," **11**:163–164

Phantom limb sensations, amputees and, **1**:73–75

Pharmacists and raised platforms, **8**:5–7

Philips screwdriver, origins of, **2**:206–207

Philtrums, purpose of, **6**:43; **8**:266–267

Photo processing, one-hour, length of black-and-white film and, **4**:39

Photographs
poses in, **6**:218
red eyes in, **4**:68–69
stars in space and, **10**:213–215

Photography
color of cameras, **7**:246
hand position of men in old, **7**:24–26
hands on chins in, **7**:203–207
Polaroid prints, flapping of, **7**:175–176
smiling in old photographs, **7**:19–23

Physical exams, back tapping during, **3**:145–146

"Pi" as geometrical term, **5**:80–81

Piano keys, number of, **10**:7–9

"Pig in a poke," **11**:25

Pigeons
baby, elusiveness of, **1**:254; **10**:253–254
loss of toes, **7**:166
whistling sound in flight, **7**:58

Pigs
curly tails of, **4**:199; **5**:218–219; **7**:228–229

Pigs (*cont.*)
   pink hair color of, **8**:98
   roasted, and apples in mouths of, **7**:159; **10**:274–275
Pilgrims, buckled hats of, **10**:47–51
Pillow tags, label warnings of, **2**:1–2
Pilots
   and dimming of interior lights, **8**:24–25
   female, dearth of, on airlines, **1**:131–133
"Pin money," **11**:115
Pine nuts, shelling of, **2**:94
Pine trees, construction sites and, **2**:147–148
Pineapple in gelatin, **3**:149–150
Pinholes, on bottle caps, **2**:223
Pink as color for baby girls, **1**:29
"Pink lady," **11**:190–191
Pink stripes on magazine labels, **8**:96–97
"Pinkie," **11**:190–191
Pins in men's dress shirts, **4**:29–30
"Pipe down," **11**:13
Pipes, kitchen, shape of, **4**:82–83
Pirates
   earrings on, **9**:43–45; **10**:272–273
   walking the plank, **9**:37–42; **10**:273–274
Pistachio ice cream, color of, **7**:12–13
Pistachios, red color of, **1**:26–28
Pita bread, pockets in, **6**:98
Pitcher's mounds
   location of, **5**:181
   rebuilding of, **9**:195–198
Pittsburgh Steelers, emblems on helmets of, **7**:67–68
Planets, twinkling of, at night, **4**:50–51

Plastic bottles, beer and, **7**:161–162
Plastic cups, shape of, **9**:289
Plastic deer ornaments on lawns, **9**:262–264
Plastics, recyclable, numbers on, **6**:155–156
Plates
   repositioning of, **8**:184; **9**:238–242
   round shape of dinner, **8**:162–164
Plots, farm, circular shape of, **7**:118–119
Plug prongs
   holes at end of, **5**:94–95
   three prongs versus two prongs, **5**:191
Plum pudding, plums in, **5**:49
Plumbing
   kitchen, shape of, **4**:82–83
   sound of running water, **3**:239–240
Pockets in pita bread, **6**:98
Poison ivy, grazing animals and, **3**:86–87
Polaroid prints, flapping of, **7**:175–176
Poles
   directions at, **10**:243
   telling time at North and South, **10**:241–243
Pole-vaulting
   preparation for different heights, **9**:97–101
   women and, **9**:102–107
Police
   and crowd estimates, **1**:250–253
   flashlight grips, **10**:30–32
   radar and speed measurement, **8**:88–91
Police car beacons, colors on, **7**:135–137

Police dogs, urination and defecation of, **3**:67–68
Policemen and mustaches, **6**:219; **7**:218–220; **8**:246–247; **9**:278
Ponds
   effect of moons on, **5**:138–139
   fish returning to dried, **3**:15–16; **10**:256
   ice formations on, **5**:82–83
   versus lakes, **5**:29–30; **7**:241
   versus lakes, level of, **9**:85–86
Poodles, wild, **6**:207–209
Pool balls, dots on, **10**:237–240
"Pop goes the weasel," **11**:14
Popcorn
   "gourmet" versus regular, **1**:176
   popping in-house, in movie theaters, **1**:45–50
   versus other corns, **3**:142–143
Popes
   name change of, **10**:17–18
   white skullcap of, **10**:80–81
   white vestments of, **10**:79–80
Popping noise of wood, in fires, **4**:10
Pork and beans, pork in, **2**:19
"Port," **11**:65–66
"Porthole," **11**:65–66
Post office, translation of foreign mail and, **3**:133
Postage and ripped stamps, **8**:62
Postage Stamps
   leftover perforations of, **4**:179
   taste of, **2**:182
Postal Service, U.S., undeliverable mail and, **5**:13–14
Pot pies, vent holes in, **6**:28
Potato chips
   bags, impossibility of opening and closing, **9**:117–118
   curvy shape, **9**:115–116
   green tinges on, **5**:136–137; **6**:275

price of, versus tortilla chips, **5**:137–138
Potato skins, restaurants and, **3**:12–13
Potatoes, baked, and steak houses, **6**:127–129
Potholes, causes of, **2**:27
"Potter's field," **11**:198
Power lines
   humming of, **9**:165–168; **10**:259
   orange balls on, **4**:18–19
Prefaces in books, versus introductions and forewords, **1**:72–73
Pregnancy, permanents and, **3**:170–171
Pregnant women, food cravings of, **10**:183–185
Preserves, contents of, **6**:140–141
Press conferences, microphones in, **2**:11–12
"Pretty kettle of fish," **11**:178
"Pretty picnic," **11**:178
Pretzels, shape of, **6**:91–92
Priests, black vestments and, **10**:77–79
Priority mail, first class versus, **3**:166–167
Prisoners and license plate manufacturing, **8**:137–139
Prohibition, liquor production of distilleries during, **9**:54–56
Pronunciation, dictionaries and, **10**:169–179
"P's and Q's," **11**:88–89
Pubic hair
   curliness of, **5**:177–178
   purpose of, **2**:146; **3**:242–243; **6**:275–276
Public buildings, temperatures in, **8**:184
Public radio, low frequency numbers of, **10**:181–183

Pudding, film on, **6**:57

Punts, measurement of, in football, **3**:124–125

Purple
Christmas tree lights, **6**:185–186; **9**:293; **10**:280
paganism, **9**:292–93
royalty and, **6**:45–46

"Put up your dukes," **11**:137

Putting, veering of ball toward ocean when, **6**:107–108

"Q.T.," **11**;59

"Qantas," spelling of, **8**:134–135

Q-Tips, origins of name, **6**:210–211

"Quack [doctor]," **11**:45

"Quarterback," **11**:138

Quarterbacks and exclamation, "hut," **6**:210

Quarter-moons versus half moons, **7**:72–73

Quarts and gallons, American versus British, **4**:114–115

Queen-size sheets, size of, **3**:87–88

Rabbit tests, death of rabbits in, **7**:69–71

Rabbits and nose wiggling, **5**:173–174

Racewalking, judging of, **9**:20–23

Racquetballs, color of, **4**:8–9

Radar and police speed detection, **8**:88–91

Radiators and placement below windows, **9**:128–130

Radio
beeps before network news, **1**:166–167
FM, odd frequency numbers of, **10**:59–60
lack of song identification, **7**:51–57

public, low frequency numbers, **10**:181–183

Radio Shack and lack of cash registers, **5**:165–166

Radios
battery drainage, **10**:259–260
lingering sound of recently unplugged, **4**:47

Railroad crossings and "EXEMPT" signs, **5**:118–119

Railroads, width of standard gauges of, **3**:157–159

Rain
butterflies and, **4**:63–64
fish biting in, **10**:131–138
measurement container for, **4**:161–163
smell of impending, **6**:170–171; **7**:241

Raincoats, dry-cleaning of, **2**:216–217

"Raining cats and dogs," **11**:26

"Raise hackles," **11**:6–7

Raisins
cereal boxes and, **2**:123
seeded grapes and, **2**:218–219

Ranchers' boots on fence posts, **2**:77–81; **3**:243–245; **4**:231; **5**:247–248; **6**:268; **7**:250; **8**:253; **9**:298–299; **10**:251–253

Razor blades, hotel slots for, **2**:113

Razors, men's versus women's, **6**:122–123

"Read the riot act," **11**:89

"Real McCoys," **11**:164–165

Rear admiral, origins of term, **5**:25

Rear-view mirrors, day/night switch on automobile, **4**:185–186

Receipts, cash register, color of, **4**:143

Records, vinyl, speeds of, **1**:58–61

Recreational vehicles and wheel covers, **7**:153

Recyclable plastics, numbers on, **6**:155–156

Recycling of newspaper ink, **7**:139–140

Red
color of beef, **8**:160–161
eyes in photographs, **4**:68–69
paint on coins, **7**:117

"Red cent," **11**:191

"Red herring," **11**:185–186

Red lights, versus green lights, on boats and airplanes, **4**:152–153

"Red tape," **11**:187

Red Wings, Detroit, octopus throwing and, **9**:183–186

"Red-letter day," **11**:186–187

Redshirting in college football, **7**:46–48

Refrigeration of opened food jars, **6**:171–172

Refrigerators
location of freezers in, **2**:230–231
smell of new, **8**:91–92
thermometers in, **10**:85–87

Relative humidity, during rain, **1**:225–226

Remote controls, men versus women and, **6**:217; **7**:193–196

Repair shops, backlogs and, **4**:45–47

Restaurants
coffee, versus home-brewed, **7**:181
vertical rulers near entrances of, **7**:95–96

Restrooms, group visits by females to, **6**:217; **7**:183–192; **8**:237–238; **9**:277–278

Revolving doors, appearance of, in big cities, **4**:171–173

Reynolds Wrap, texture of two sides of, **2**:102

Rhode Island, origins of name, **5**:21–22

Ribbons, blue, **3**:57–58

Rice cakes, structural integrity of, **10**:9–11

Rice Krispies
noises of, **3**:165; **5**:244–245
profession of Snap!, **8**:2–4

"Right wing," **11**:116

"Rigmarole," **11**:81–82

Rings, nose, bulls and, **10**:147–148

Rinks, ice, temperature of resurfacing water in, **10**:196–198

"Riot Act [1716]," **11**:89

Roaches
automobiles and, **7**:3–4; **8**:256–257; **9**:298
position of dead, **3**:133–134; **8**:256
reactions to light, **6**:20–21

Roads
blacktop, coloring of, **5**:22–23
fastening of lane reflectors on, **5**:98–99
versus bridges, in freezing characteristics, **5**:193

Robes, black, and judges, **6**:190–192

Rocking in zoo animals, **10**:279

Rodents and water sippers, **6**:187

Roller skating rinks, music in, **2**:107–108; **6**:274–275

Rolls
coldness of airline, **3**:52–53
versus buns, **8**:65–66

Roman chariots, flimsiness of, **10**:105–107

Roman numerals
calculations with, **3**:105–106
copyright notices in movie credits and, **1**:214–216
on clocks, **5**:237–238

Roofs, gravel on, **6**:153–154
Roosevelt, Teddy, and San Juan
 Hill horses, **4**:49
Roosters, crowing and, **3**:3
Root beer
 carbonation in, **1**:87–88
 foam of, **8**:93–94
Rubble, Betty
 nonappearance in Flintstones
  vitamins, **6**:4–5; **9**:285–286
 vocation of, **7**:173–174
Ruins, layers of, **2**:138–140
Rulers, vertical, in restaurant en-
  trances, **7**:95–96
Run amok, **11**:68
Runny noses
 cold weather and, **10**:146–147
 kids versus adults, **9**:89–90
Rust, dental fillings and, **10**:41–42
RVs and wheel covers, **7**:153
"Rx," **11**:59

S.O.S Pads, origins of, **8**:103–
  104
"Sacked [fired]," **11**:75
Sacks, paper
 jagged edges on, **6**:117–118
 names on, **2**:166–167
Safety caps, aspirin, 100–count
  bottles of, **4**:62
Safety pins, gold versus silver,
  9:87–88
Saffron, expense of, **1**:129–130
Sailors, bell bottom trousers and,
  2:84–85
"Salad days," **11**:178
Salads, restaurant, celery in,
  6:218; **7**:207–209
Saloon doors in Old West, **3**:198
Salt
 and pepper, as table condiments, **5**:201; **6**:225–226;
  8:235–236
 in oceans, **5**:149–150

packaged, sugar as ingredient
 in, **4**:99
round containers and, **10**:149–
 150
storage bins on highway and,
 10:216–218
versus sand, to treat icy roads,
 2:12–13
Salutes, military, origins of, **3**:147–
 149
San Francisco, sourdough bread
 in, **2**:180–181
Sand
 in pockets of new jeans, **7**:152
 storage bins on highway and,
  10:216–218
 versus salt, to treat icy roads,
  2:12–13
Sandbags, disposal of, **10**:193–195
Sardines, fresh, nonexistence in
  supermarkets, **1**:76–78
"Sawbuck [ten-dollar bill],"
 11:119–120
Sawdust on floor of bars, **10**:118–
 120
Scabs, itchiness of, **5**:125–126
Scars, hair growth and, **2**:186
"Schneider," **11**:136
School clocks, backward clicking
 of minute hands in, **1**:178–
 179
Schools, CPR training in, **7**:31–
 33
Scissors, sewing, and paper cut-
 ting, **8**:131–132
"Scot-free," **11**:143
"Scotland Yard," **11**:144
Scotsmen and kilts, **7**:109–110
Screen doors, location of handles
 on, **7**:91
Screwdrivers, reasons for Philips,
 2:206–207
Scuba masks, spitting into, **9**:30–
 34

Sea gulls in parking lots, **6**:198–199; **10**:254–256

Sea level, measurement of, **4**:154–155

Seas versus oceans, differences between, **5**:30–32

Seat belts
  and shoulder straps in airplanes, **8**:141–142
  in buses, **1**:84–85
  in taxicabs, **1**:85

Secretary as U.S. government department head designation, **3**:121–122

"Seed [tournament ranking]," **11**:141–142

"Seeing stars," head injuries and, **10**:156–158

Self-service versus full-service, pricing of, at gas stations, **1**:203–209

"Semi," origins of term, **2**:179

"Semimonthly," **11**:194

"Semiweekly," **11**:194

Serrated knives, lack of, in place settings, **4**:109–110

Settling in houses, **6**:32–34

Seven-layer cakes and missing layers, **6**:80–81

Seventy-two degrees, human comfort at, **2**:178–179

Sewing scissors and paper cutting, **8**:131–132

Shampoo bottles, notches on bottom of, **10**:29–30

Shampoo labels, "FD&C" on label of, **4**:163

Shampoos
  colored, white suds and, **4**:132–133
  lathering of, **5**:44–45
  number of applications, **1**:90–93

Shaving of armpits, women and, **2**:239; **3**:226–229; **6**:249

Sheets
  irregular, proliferation of, **1**:145–147
  queen-size, size of, **3**:87–88

Sheriffs' badges, shape of, **5**:73–74

Shirts
  buttons on men's versus women's, **2**:237–238; **3**:207–209; **5**:226; **7**:223
  men's, pins in, **4**:29–30
  single-needle stitching in, **6**:51
  starch on, **3**:118

Shoe laces
  length in athletic shoes, **8**:41–42
  untied, in shoe stores, **8**:40

Shoe sizes, differences between, **1**:65–70

Shoes
  lace length in shoe stores, **8**:41–42
  layers on, **5**:59
  of deceased, at funerals, **7**:153–154
  penny loafers, **8**:43–44
  single, on side of road, **2**:236–237; **3**:201–207; **4**:232–233; **5**:225–226; **6**:245–248; **7**:221–222; **8**:228–230; **9**:271–272
  tied to autos, at weddings, **1**:235–238
  uncomfortable, and women, **1**:62–64
  untied laces in stores, **8**:40
  wing-tip, holes in, **8**:44

"Shoofly pie," **11**:179

Shopping, female proclivity toward, **7**:180; **8**:205–209

Shopping malls, doors at entrance of, **6**:180–181

"Short shrift," **11**:15

Shoulder straps and seat belts in airplanes, **8**:141–142

Shredded Wheat packages, Niagara Falls on, **5**:100–101
"Shrift," **11**:15
Shrimp, baby, peeling and cleaning of, **5**:127
"Shrive," **11**:15
"Siamese twins," **11**:151
"Sic," as dog command, **5**:51
Side vents in automobile windows, **6**:13–14
"Sideburns," **11**:126–127
Sidewalks
 cracks on, **3**:176–178
 glitter on, **7**:160; **10**:61–62
Silica gel packs in electronics boxes, **6**:201
Silos, round shape of, **3**:73–74; **5**:245; **10**:260–261
Silver fillings, rusting of, **10**:41–42
Silverstone, versus Teflon, **2**:3
Singers, American accents of foreign, **4**:125–126
Single-needle stitching in shirts, **6**:51
Sinks, overflow mechanisms on, **2**:214–215
Siskel, Gene, versus Roger Ebert, billing of, **1**:137–139
Skating music, roller rinks and, **2**:107–108; **6**:274–275
Skating, figure, and dizziness, **5**:33–35
Ski poles, downhill, **3**:69
"Skidoo," **11**:100
Skunks, smell of, **10**:88–91
Skyscrapers, bricks in, **6**:102–103
Skytyping, versus skywriting, **9**:17–18
Skywriting
 techniques of, **9**:12–16
 versus skytyping, **9**:17–18
Sleep
 babies and, **6**:56–57

drowsiness after meals, **6**:138–139
 eye position, **6**:146
 heat and effect on, **6**:137–138
 twitching during, **2**:67
"Slippery When Wet" signs, location of, **7**:167–168
"Small fry," **11**:179
Smell of impending rain, **6**:170–171; **7**:241
Smiling in old photographs, **7**:19–23
Smoke from soda bottles, **8**:148
Snack foods and prepricing, **3**:79–80
Snake emblems on ambulances, **6**:144–145; **7**:239–240
Snakes
 sneezing, **7**:98
 tongues, **2**:106; **10**:278
Snap! [of Rice Krispies], profession of, **8**:2–4
"Snap! Crackle! And Pop!" of Rice Krispies, **3**:165
Sneezing
 eye closure during, **3**:84–85
 looking up while, **2**:238
 pepper and, **8**:61
 snakes and, **7**:98
Snickers, wavy marks on bottom of, **6**:29–31
Sniffing, boxers and, **2**:22–23; **10**:256–258
Snoring, age differences and, **10**:126–127
Snow and cold weather, **3**:38
"Snow" on television, **9**:199–200
Soap, Ivory, purity of, **2**:46
Soaping of retail windows, **9**:265–267
Soaps, colored, white suds and, **4**:132–133
Social Security cards, lamination of, **5**:140–141

MASTER INDEX OF IMPONDERABILITY

Social Security numbers
  fifth digit of, **8**:100–102
  reassignment of, **5**:61–62
  sequence of, **3**:91–92; **6**:267
Socks
  angle of, **3**:114–115
  disappearance of, **4**:127–128;
    5:245–246; **6**:272; **8**:266
  men's, coloring of toes on,
    4:19–20
"Soda jerk," **11**:176
Soft drinks
  bottles, holes on bottom of,
    6:187–188
  brominated vegetable oil in,
    6:96–97
  calorie constituents, **6**:94–95
  effect of container sizes and
    taste, **2**:157–159
  filling of bottles of, **4**:53
  finger as fizziness reduction
    agent, **9**:28–29
  fizziness in plastic cups, **9**:26
  fizziness of soda with ice cream,
    9:27
  fizziness over ice, **9**:24–25
  freezing of, in machines, **9**:10–
    11
  Kool-Aid and metal containers,
    8:51
  machines, "Use Correct
    Change" light on, **9**:186–188
  phenylalanine as ingredient in,
    6:96–97
  pinholes on bottle caps of,
    2:223
  root beer, foam of, **8**:93–94
  smoke of, **8**:148
  "sodium-free" labels, **4**:87–88
Soles, sunburn and, **8**:63–64
"Son of a gun," **11**:82
Sonic booms, frequency of, **4**:23
Sororities, Greek names of,
  10:94–98

Souffles and reaction to loud
  noises, **7**:41–42
Soup, alphabet, foreign countries
  and, **10**:73
Soups and shelving in supermarkets, **6**:26–27
Sour cream, expiration date on,
  3:132
Sourdough bread, San Francisco,
  taste of, **2**:180–181
South Florida, University of, loca-
  tion of, **4**:7–8
South Pole
  directions at, **10**:243
  telling time at, **10**:241–243
Sparkling Wine, name of, versus
  champagne, **1**:232–234
Speech, elderly versus younger
  and, **6**:24–25
Speed limit, 55 mph, reasons for,
  2:143
Speeding and radar, **8**:88–91
Speedometers, markings of, in au-
  tomobiles, **2**:144–145
Spelling, "i" before "e" in, **6**:219;
  7:209; **8**:240–245
Sperm whales, head oil of, **6**:87–
  89
"Spic and span," **11**:41
Spiders and web tangling, **7**:169–
  170
Spitting, men versus women,
  7:181–182; **8**:226–227
Spoons, measuring, inaccuracy of,
  1:106–107
Sprinkles, jimmies versus,
  10:165–168
Squeaking, causes of, **9**:205–
  207
"St. Martin's Day," **11**:146
Stage hypnotists, techniques of,
  1:180–191
Staining, paperback books and,
  2:93–94

Stains, elimination of, on clothing, **6**:77–78

Staling of bread, **7**:125–126

Stamp pads, moisture retention of, **3**:24

Stamps
perforation remnants, **4**:179
postage, taste of, **2**:182
validity of ripped, **8**:62

Staplers
fitting of staples into, **10**:187–189
outward setting of, **10**:189–190

Staples
clumping of, **10**:46
fitting into staplers of, **10**:187–189

"Starboard," **11**:65–66

Starch on shirts, **3**:118

Stars in space, photos of, **10**:213–215

Starving children and bloated stomachs, **7**:149–150

States, versus commonwealths, **7**:119–121

Static electricity, variability in amounts of, **4**:105–106

Steak houses and baked potatoes, **6**:127–129

Steak, "New York," origins of, **7**:155–156; **8**:252

Steam and streets of New York City, **5**:16–17

Steelers, Pittsburgh, helmet emblems of, **7**:67–68

Steins, beer, lids and, **9**:95–96

Stickers, colored, on envelopes, **4**:83–84

Stickiness of peanut butter, **10**:204–207

Stock prices as quoted in eighths of a dollar, **3**:112–113

Stocking runs
direction of, **2**:124–125

effect of freezing upon, **2**:125–126

"Stolen thunder," **11**:120

Stomach, growling, causes of, **4**:120–121

Stomachs, bloated, in starving children, **8**:254

"STOP" in telegrams, **3**:76–77

Strait, George, and hats, **6**:274

Straws, rising and sinking of, in glasses, **5**:36–37

Street addresses, half-numbers in, **8**:253

Street names, choice of, at corner intersections, **1**:154–156

Street-name signs, location of, at intersections, **1**:136

Streets, glitter and, **7**:160; **10**:61–62

String cheese, characteristics of, **3**:155

String players
left handed, in orchestras, **10**:276–277
left-handed, in orchestras, **9**:108–109

Styrofoam coolers, blue specks on, **10**:130–131

Submarines, anchors and, **4**:40–41

Sugar
clumping together of, **3**:103–104
spoilage of, **2**:85

Sugar cube wrappers, slits in, **1**:170–171

Sugar Frosted Flakes, calorie count of, **1**:38–40

Sugar in packaged salt, **4**:99

Summer, first day of, **3**:139–141

Sunburn
delayed reaction in, **7**:114–116
palms and soles, **8**:63–64

"Sundae," **11**:181

Tiles, ceramic, in tunnels, **6**:135–136; **8**:257–258

"Tinker's dam," **11**:82–83

Tinnitus, causes of, **2**:115–116

Tips, waiter, and credit cards, **8**:263

Tiredness and eye-rubbing, **10**:103–105

Tires
atop mobile homes, **6**:163–164
automobile tread, disappearance of, **2**:72–74
bicycle, **2**:224–226
bluish tinge on new whitewalls, **6**:192–193
inflation of, and gasoline mileage, **6**:193–195
white wall, size of, **2**:149

Tissue paper in wedding invitations, **4**:116–117

Title pages, dual, in books, **1**:139–141

Toads, warts and, **10**:121–123

"Toady," **11**:83

"Toast [salute]," **11**:182

Toasters, one slice slot on, **3**:183–185

Toenails, growth of, **3**:123

Toffee, versus caramels, **1**:64

Toilet paper, folding over in hotel bathrooms, **3**:4

Toilet seats in public restrooms, **3**:137–138

Toilets
flush handles on, location of, **2**:195–196
loud flushes of, in public restrooms, **4**:187
seat covers for, **2**:83

"Tom Collins [drink]," **11**:165

"Tommy gun," **11**:166

Tongues, sticking out of, **3**:199; **7**:224–225

Tonsils, function of, **5**:152–153

Toothpaste
expiration dates on, **4**:169–170; **6**:266
taste of, with orange juice, **10**:244–246

Toothpicks
flat versus round, **1**:224–225; **3**:237–238
mint flavoring on, **4**:153

Tootsie Roll Pops, Indian and star on wrappers of, **8**:37–38

Toques, purpose of, **3**:66–67

Tornadoes
lull before, **8**:49–50
trailer parks and, **1**:101

Tortilla chips
black specks on, **10**:23–25
price of, versus potato chips, **5**:137–138

Tortillas
black specks on, **10**:23–25
size of, versus corn, **10**:142–145

Touch tone telephones, keypad configuration for, **2**:14–15

Towels
number of, in hotel rooms, **4**:56–57
smell of used, **3**:24
textures of, **9**:79–80

Traffic control
55 mph speed limit, **2**:143
traffic lights, **2**:126–127

Traffic flow, clumping of, on highways, **4**:165–167

Traffic jams, clearing of, on highways, **1**:25–26; **8**:257

Traffic lights
colors on, **2**:126–127; **5**:246
timing of, in New York and Washington D.C., **1**:109–112

Traffic noise in U.S., versus other countries, **5**:213–214; **6**:257–258

Traffic signal light bulbs, lifespan of, **3**:31–32

Traffic signs
"DONT WALK," lack of apostrophe on, **6**:75–76
"FALLING ROCK," purpose of, **6**:72–74
placement of "Dangerous Curve," **2**:119–120

Trailer parks
tires atop mobile homes in, **6**:163–164
tornadoes and, **1**:101

Train stations, ceilings of, **8**:66–67

Trains
backwards locomotives in, **5**:15–16
"EXEMPT" signs at railroad crossings, **5**:118–119

Tread, tire, disappearance of, **2**:72–74

Treasurer, U.S., gender of, **8**:45–46

Treasury, printing of new bills by, **3**:126–128

Treble clefs, dots on, **10**:210–213

Trees
bark, color of, **6**:78–79
growth in cities, **8**:78–80
growth on slopes of, **6**:141–142; **7**:244

Triton, orbit pattern of, **4**:117–118

Tropical fish, oxygen in, **7**:84–85

Trucks
idling of engines of, **7**:150–151
license plates on, **3**:96; **10**:270–271
origins of term "semi" and, **2**:179

"Tsk-Tsk," stroking of index fingers and, **4**:198

Tuba bells, orientation of, **8**:147–148

Tuesday
release of CDs, **10**:108–112
U.S. elections and, **1**:52–54

Tumbleweed, tumbling of, **6**:5–7

Tuna, cat food, cans of, **8**:8–10

Tunnels, ceramic tiles in, **6**:135–136; **8**:257–258

Tupperware and home parties, **3**:25–26

"Turkey," in bowling, origin of term, **4**:48–49; **6**:269

Turkeys
beards on, **3**:99
white versus dark meat, **3**:53–54

Turn signals in automobiles, clicking sounds of, **6**:203

TV Guide, order of listings in, **4**:91–92

Twenty-four second clock in NBA basketball, **1**:29–31

Twenty-one as age of majority, **7**:50–51

Twenty-one gun salute, origins of, **2**:68–70

"Twenty-three skidoo," **11**:100

Twins, identical, DNA and, **10**:18–20

Twitches during sleep, **2**:67

TWIX cookie bars, holes in, **6**:28–29

"Two bits," origins of term, **2**:191–192

Two by fours, measurement of, **2**:87–88

Two-minute warning, football and, **10**:150–151

Typewriter keys, location of, **1**:127–128

"U" in University of Miami football helmets, **8**:171–172

Ultraviolet and attraction of insects, **8**:158–159

Wrapping
    Burger King sandwiches,
        8:112–113
    chewing gum, **8**:111–112
    gift box chocolate, **8**:122–123
Wrinkles on extremities, **2**:112
Wrists as perfume target, **6**:90
Wristwatches, placement on left
        hand, **4**:134–135; **6**:271

X
    as symbol for kiss, **1**:128–129
    as symbol in algebra, **9**:131–
        132
"X ray," **11**:49–50
"Xmas," origins of term, **2**:75
X-rated movies, versus XXX-rated
        movies, **2**:141–142
"XXX [liquor]," **11**:58

Yawning, contagiousness of, **2**:238;
        3:213–217; **8**:231–232
Yeast in bread, effect of, **3**:144–
        145

Yellow, aversion of insects to,
        8:158–159
Yellow Freight Systems, orange
        trucks of, **6**:64–65
Yellow lights, timing of, in traffic
        lights, **1**:109–112
Yellow Pages, advertisements in,
        3:60–63
Yellowing of fingernails, nail polish
        and, **7**:129–130
YKK on zippers, **5**:180
Yogurt
    fruit on bottom, **7**:83
    liquid atop, **7**:82

Zebras and riding by humans,
        8:139–141
ZIP code, addresses on envelopes
        and, **3**:44
"Zipper," **11**:129
Zippers, YKK on, **5**:180
Zodiac, different dates for signs
        in, **4**:27–28
Zoo animals, rocking in, **10**:279

Help!

We hate to end the book on a downbeat note, but we have to admit one dread fact: Imponderability is not yet smitten. Let's stamp it out.

"How?" you ask. Send us letters with your Imponderables, answers to Frustables, gushes of praise, and even your condemnations and corrections.

Join your inspired comrades and become a part of the wonderful world of *Imponderables*. If you are the first person to submit an Imponderable we use in the next volume, we'll send you a complimentary copy, along with an acknowledgment in the book.

Although we accept "snail mail," we strongly encourage you to e-mail us if possible. Because of the volume of mail, we can't always provide a personal response to every letter, but we'll try—a self-addressed stamped envelope doesn't hurt. We're much better with answering e-mail, although we fall far behind sometimes when work beckons.

Come visit us online at the *Imponderables* website, where you can pose Imponderables, read our blog, and find out what's happening at Imponderables Central. Send your correspondence, along with your name, address, and (optional) phone number to:

feldman@imponderables.com
http://www.imponderables.com

Imponderables
P.O. Box 116
Planetarium Station
New York, NY 10024-0116

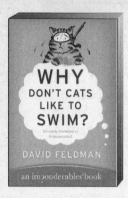
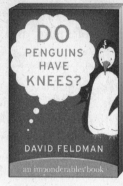